The Italian Utilities Industry

Andrea Gilardoni
Editor

The Italian Utilities Industry

Success Stories and Future Perspectives

Springer

Editor
Andrea Gilardoni
AGICI Finanza d'Impresa
Milan, Italy

ISBN 978-3-030-37676-5 ISBN 978-3-030-37677-2 (eBook)
https://doi.org/10.1007/978-3-030-37677-2

This Springer imprint is published by the registered company Springer Nature Switzerland AG
The registered company address is: Gewerbestrasse 11, 6330 Cham, Switzerland

Preface

The energy industry in the world and in Italy has gone through a radical transformation in the last twenty years, and the process is still ongoing. Changes in the regulatory framework, technological evolution and increasing concern for the environment have revolutionised the sector structure and forced Utilities to adapt their organizations and business models. European directives have driven this revolution, firstly by breaking the existing monopolies in the relevant markets (power, gas, waste) and secondly by paving the way for the decarbonisation of the industry with ambitious targets. Today, a new, even harder, challenge has risen for Utilities: the so-called energy transition which is redesigning the whole energy landscape. A combination between digitalisation, spread of renewables and distributed generation, has given birth to a new paradigm, to which Utilities are adapting according to their characteristics.

This book sheds light on the main dynamics of the energy and environmental sectors occurred in the past two decades in Italy, and to some extent in Europe, based on the direct experiences of the leading players. We have collected experiences from the most important Utilities, financial players, consultants and regulator, to give an in-depth and comprehensive overview of the Italian landscape from a privileged perspective.

Since the book is intended as a joint effort of the entire sector, we gave voice directly to renowned experts who have had an influential role in the past decades. Therefore, each chapter was drafted by top representatives of the most important organizations, choosing the topic they considered the most significant in shaping the sector. The heterogeneity of the contributions reflects the different points of view of the various authors in an enriching polyphonic perspective.

For all these reasons, I want to thank all the authors and their respective teams, for the invaluable participation: Nicola Monti (Edison S.p.A., Chief Executive Officer), Francesco Starace (Enel S.p.A, Chief Executive Officer), Alessandro Garrone (ERG S.p.A, Executive Vice President), Giovanni Valotti (A2A, Chairman), Stefano Venier, Stefano Verde (respectively Chief Executive Officer and Strategic Planning and Policy Making, HERA S.p.A.), Massimiliano Bianco (IREN S.p.A., Chief Executive Officer), Enrico De Girolamo (CVA S.p.A., Chief Executive Officer),

PierPaolo Carini, Rosario Bisbiglia (respectively Chief Executive Officer, Egea S.p.A., and Managing Director, Egea Commerciale Srl), Luigi Ferraris (Terna S.p.A., Chief Executive Officer), Livio Gallo and Vincenzo Ranieri (respectively Head of Enel Global Infrastructure and Networks and Chief Executive Officer of E-Distribuzione S.p.A.), Marco Alverà (Snam S.p.A., Chief Executive Officer), Paolo Gallo (Italgas S.p.A, Chief Executive Officer), Luca D'Agnese (Cassa Depositi e Prestiti S.p.A., Chief Infrastructure & Public Sector Officer), Luca Matrone, Matteo Balasso, Urbano Maria Cremona (respectively Global Head of Energy and Business Analysts, Intesa Sanpaolo S.p.A.), Pierfederico Pelotti, (Accenture Italy, Head of Utilities, Resources Area), and Guido Bortoni (Former President of ARERA). I would also like to thank ACEA S.p.A. for its contribution to the publication of this volume.

Last but not least, a special thank is for Prof. Michele D'Alessandro (Bocconi University), Marco Carta, and Michele Perotti (respectively Chief Executive Officer and Senior Analyst, Agici Finanza d'Impresa) and to the whole Agici team whose contribution has been essential for ideas and comments, and also to give to this book its final shape.

Finally, I would like to dedicate this effort to my generous and intelligent wife and to all my grandchildren that I hope will face their life with courage, honesty and love.

Milan, Italy
February 2020

Andrea Gilardoni

Contents

Abbreviations

ARERA	Autorità di Regolazione per Energia Reti Ambiente
AS	Ancillary services
ATR	Autothermal reforming
Bcm	Billion cubic metres
CAGR	Compound annual growth rate
CCGT	Combined cycle gas turbine
CCS	Carbon capture and storage
CCUS	Carbon capture, utilisation and storage
C&D	Construction and demolition
CSR	Corporate social responsibility
CSR	Corporate social responsibility
CSV	Creating shared value
DER	Distributed energy resources
DR	Demand response
DSM	Demand side management
DSO	Distribution system operator
EE	Energy efficiency
ERM	Enterprise risk management
ESCO	Energy service company
ESG	Environmental, social, and governance
ET	Energy transition
EV	Electric vehicle
G2P	Gas-to-power
GBS	Gravity-based structure
Gcm	Giga cubic metres
GIS	Geographic information system
GME	Gestore dei Mercati Energetici
GSE	Gestore dei Servizi Energetici
GWh	Gigawatt hours
IoT	Internet of things

LCOE	Levelized cost of electricity
LNG	Liquefied natural gas
Mcm	Million cubic metres
MS	Member state
MSR	Methane steam reforming
O&M	Operations and maintenance
OECD	Organisation for Economic Co-operation and Development
P2G	Power-to-gas
PAN	Puglia Active Network
PEM	Polymer electrolyte membrane
PNIEC	Piano Nazionale Integrato per l'Energia e il Clima
POD	Point of delivery
PPA	Power purchase agreement
PV	Photovoltaic
RES	Renewable energy sources
SDG	Sustainable development goal
SEN	Strategia energetica nazionale
SMR	Steam methane reforming
SOEC	Solid oxide electrolyser cell
SPV	Special-purpose vehicle
TPA	Third-party access
TSO	Transmission system operator
TWh	Terawatt hours
UN	United Nations
UNEP	United Nations Environment Programme
UNFCCC	United Nations Framework Convention on Climate Change
V2G	Vehicle to grid

Part I
2000–2020 A Revolution in the Italian Utilities Industry: What's Next?

Utilities: Fundamental Players for Socio-economic and Environmental Development

Andrea Gilardoni

Abstract The environment in which Utilities operate has dramatically evolved along the past twenty years and the transformation is still ongoing. Changes in the regulatory framework, technological evolution, increasing concern for the environment, have revolutionised the sector structure and forced Utilities to adapt their organizations and business models. European directives have been a fundamental driver of this transformation, firstly by breaking the existing monopolies in the relevant markets (power, gas, waste) and secondly by establishing ambitious decarbonisation targets. Moreover, shocks in the economy as the long economic crises begun in 2008, which affected power demand, have been another key element for companies' strategy. Recently, the combination between digitalisation, spread of renewables and distributed generation, has given birth to the so-called energy transition, a new paradigm in the energy sector to which Utilities are adapting. This introductive chapter clearly identifies these and further major steps in the transformation of the Utilities sector and the most significant trends that have shaped the companies' strategic choices, which will be discussed in depth in the subsequent contributions realized by leading experts.

Keywords Utilities · Liberalisation · Aggregation processes · Business transformation · RES · Energy Transition

1 Brief Profile of the Utilities Sector

The aim of this book is to discuss the main dynamics of the energy and environmental sectors—i.e. Power, Gas, Water and Waste—in the 20 years between 2001 and 2020 in Italy and—to some extent—in Europe, based on the experiences of a number of leading companies and organizations that collaborated with the Observatory on Alliances and Strategies in the pan-European Utility Market[1] (in the following pages, the Observatory). The objective is to shed light on the strategies

---

[1]The Observatory, led by Agici Finanza d'Impresa in collaboration with Accenture, has reached in 2020 its 20th edition. See Appendix for the complete list of publications from the Observatory.

A. Gilardoni (ed.), *The Italian Utilities Industry*,
https://doi.org/10.1007/978-3-030-37677-2_1

adopted in the industry in a period of radical change that characterized the last two decades. This not only for the sake of a better historical understanding, but for designing the future of this industry where the challenges are still numerous and relevant.

In several European countries, starting from the industrial revolution of the 18th century, we have witnessed, at the national and local level, the development of companies providing essential services, which were fundamental for the socio-economic and environmental development of Countries, Regions or Cities. There are many examples of these services: from the slaughterhouses to milk provision, from power generation and distribution to water supply and treatment, from natural gas distribution to waste collection and disposal, from public transportation to telecommunication. The degree of essentiality is geographically and historically dependent: a certain service may be key for a developing country while being less relevant in an advanced nation. Similarly, other services that are essential in a certain period can become irrelevant over time.

With the aim of better tracing the profiles of the sectors considered, some reflections are appropriate.

Industrial profile. The sectors considered have characteristics that imply high levels of competencies, especially those developed in the last few years. Part of the supplies are located in plants where technological, financial, and organizational problems require an approach that is typical of manufacturing companies; incinerators, wastewater treatment plants, power plants are good examples.

Grid. To provide services, the grids and networks are indispensable components. Developing this infrastructure is important to reach consumers, that can be several millions for the leading companies. Again technologies, investments and organization are vital for quality and timeliness.

Monopolistic position. Most of the companies in the sectors investigated used to be (and still are) in a monopolistic position, with all the related implications in terms of low efficiency. Most of the European directives that we discuss later were aimed at eliminating the most negative aspects of this market structure.

Public ownership. A large proportion of companies in the sectors investigated are controlled or wholly owned by public administrations such as municipalities, regions or central governments. Of course, this has significant implications on governance for fundamental decisions and on management and efficiency.

Environmental challenges. In the last decade, the environment has become a key topic. Our companies are at the forefront to face climate change; this for emission reasons (i.e. power generation), for sanitary motivations (i.e. water supply), for pollution prevention (e.g. waste management), and for the construction of essential infrastructures (i.e. Snam and Terna). As we shall see, many of the European directives have put environmental issues at the center of interest.

In this chapter, we will discuss the main trends, on the industrial and policy ground, which shaped the Utilities sector in Italy and in Europe in the last two decades, as today the framework in which companies operate is increasingly dependent on UE regulation. Our analysis will consider all the public services typically distributed by Utilities, but will concentrate on the energy-related businesses since these are the most affected by change in regulation, technologies, and market structure.

2 The European Directives as a Driver of Change

Most of the discussions in these introductive pages are based on research performed by the mentioned Observatory. It was created in 2000 with the objective of analyzing the strategic responses of leading Italian companies (and later also continental companies[2]) to the European Union directives on Power, Gas, Waste and Water transposed by the Italian legislator to national law in the last decade of the 20th century. In particular, we refer to the following Directives:

(1) **Waste**: 91/156/EEC of 18th March 1991; 91/689/EEC of 12th December 1991; 94/62/EC of 20th December 1994; adopted in Italy on the 5th February 1997, n.22 (Ronchi Decree)
(2) **Power**: 96/92/EC of 19th December 1996, adopted in Italy on the 16th March 1999, n.79 (Bersani Decree)
(3) **Gas**: 98/30/EC of 28th June 1998, adopted in Italy on the 23rd May 2000, n.164 (Letta Decree).

In addition, one should consider the 5th of January 1994, n. 36 law (Legge Galli) aimed at remodeling the water industry in a more integrated perspective. This law, too, was based on several European directives.

Which were the objectives of the European directives? Why did Europe decide to intervene so heavily on these topics? If we refer to power, the rationale was to create a level playing field that facilitated the integration of the national markets, towards the creation of a European energy market. This, in turn, would enable the achievement of further objectives:

(1) To increase the low level of competition, which was a reason of inefficiency, low modernization, high services cost and social inequality.
(2) To ensure security of energy supply, through the diversification of the fuel mix and supply sources and increasing the resilience of the energy networks.
(3) To facilitate the access of energy products to remote and less-developed regions, in accordance with the EU objective to reduce the gap among rich and poor areas in the Union.

[2]The Observatory started with a focus on the Italian industry but after a few years extended its attention to Europe, sometimes including global companies too.

(4) To guarantee a sustainable development of the utilities sector, in view of protecting the environment and fighting climate change.
(5) To spur investments. A stable and common regulatory framework is key to stimulate investments, especially large cross-border projects.
(6) To promote technological innovation. R&D in this field can require large amounts of resources (e.g. on nuclear), which are easier to provide thanks to international cooperation. Moreover, an integrated market would facilitate the dissemination of knowledge and technology.

These directives, although different in many aspects, called for a cultural revolution in all the considered industries: the monopolies which had characterized the sectors until that moment were to be broken. This led to competition, development of more transparency, marketing competencies, more attention to customers, and many more benefits for consumers and companies.

The magic word was Liberalization. It meant above all giving the chance to each final consumer to choose his/her supplier of gas and power, forcing companies into a competitive setting. To make this possible, it was essential to develop an unbundling process i.e. to separate the transportation and distribution networks (regulated businesses) from the generation and selling activities (non-regulated businesses). A related concept was the Third-Party Access right, i.e. the possibility for suppliers (given certain technical prerequisites) to provide services to each client in the national territory.

To understand the level of change in Italy (but this is true in most countries of the world), we should consider that the organizations[3] providing services in the selected sectors were—and are—very often totally owned or controlled by Local, Regional or National administrations; the monopolistic and protected environment was expected to disappear. And besides, the geographic boundaries, historically defined by the Administrations' responsibility, could be penetrated by new entrants or crossed through in order to extend the geographic scope.

To make the picture certainly more complex, in those years a number of additional laws were seriously discussed or approved. Examples are those concerning the redesign of the role of Local Administrations regarding the public services, the ones concerning local transportation or telecommunication, industries considered being among the public utilities. These legislative dynamics, that continued for several decades (and that for certain aspects are still ongoing) put companies in a never-ending uncertainty as regards the competitive setting of the sector and their possible and desirable strategic moves.

[3]We use the word organizations because very often the services were offered directly by the Public Administration—like the Municipality, the Region or the State—through bodies such as agencies and public entities. For instance, Enel became a joint-stock company only in 1992.

3 The Introduction of Regulation and the Role of a Newly Created Authority

A crucial role in the modernization of the energy and utilities markets has been played by the Regulatory Authority for Energy, Network and Environment: ARERA (*Autorità di Regolazione per l'Energia, le Reti e l'Ambiente*). Since its establishment with law n.485 of 14th November 1995, it has had the mandate of protecting consumers' interests and promoting competition, efficiency and high-quality service standards.

The regulation and control responsibility of the Authority, initially limited to the power and gas sectors, was extended to the following public services by the related regulatory provisions:

- Water services, with law n.214/2011;
- District heating and cooling, with law n.102/2014 which adopted the European directive 2012/27/EU on energy efficiency (EE), included;
- Waste collection and recycling, with law n.205/2017;

The Authority's powers are the same in all the mentioned industries, and they include control, inspection, sanctioning. Another fundamental task of ARERA is to establish the methodology for the calculation of tariffs for the regulated activities, which has a direct impact on revenues and investments.

Even if it was founded some years before, the Authority's role became much more important after the Bersani decree (1999), which started the privatization and liberalization process of the energy sector. With Enel's unbundling and privatization, the State no longer had direct control on power generation, distribution and sales, thus a regulatory Authority was necessary to guarantee consumers' protection and the quality of service. The same reasoning applies to the gas market, where the liberalization process started with the Letta decree (2000).

Today, the Authority's activity in the regulated sectors goes beyond the promotion of competition and efficiency of the services, and has the aim to:

- Ensure homogeneous availability and dissemination of the services on the whole national territory;
- Define adequate quality standards for the provided services;
- Arrange a stable tariff system based on transparent criteria;
- Safeguard the interests of consumers.

These functions are carried out by means of harmonizing the economic targets of the operators with the general targets of the community, on a social and environmental ground and in view of an efficient use of resources.

As for its future activity, the Authority's objectives will revolve around the themes of decarbonization, circularity and mitigation of climate change effects (especially those linked to drought). To this aim, ARERA will need to spur investments in the following fields: securing of water sources; completion of the sewerage; improvement of the purification capacity; evolution of the electric transmission and distribution networks, fundamental to include an increasing share of RES and distributed generation.[4]

4 The Aggregation/Concentration Processes

The mentioned European directives were perceived by most of the utilities as a strong threat even to their survival. On one hand, the competition in other territories was something totally new for most of them; on the other, the expected liberalization of the Italian market was a strong attraction for foreign players who became equipped to penetrate it. Thames Water, E.On, ENBW, Suez Lyonnaise des Eaux, Electrabel, Iberdrola, Enron, Foster-Wheeler, Marubeni Power, were just examples of international players interested in a country like Italy, relatively advanced, with a good growth potential, with needs of renovating/extending the infrastructure and with a clear privatization program. The creation of an independent Authority was another guarantee for foreign investors in power and gas.

What were the reactions of the Italian players? Probably, the dimensional issue became the first and most relevant component of any strategy to face the threats and specially to defuse the potential invasion of large foreign players. In other words, the Italian industry realized that it was too fragmented, as only few players were able to deploy the resources to face the emerging competition. The need to grow dimensionally became evident and widespread in the opinion of researchers and politicians with a broad vision. The following are only some of the most relevant benefits: stronger bargaining power with suppliers and third parties in general, economies of scale concerning plants and networks, greater efficiency in facing the changes implied by the new competition, stronger financial capabilities at lower interest rates, larger resources (financial and human) to expand in new markets, larger returns on investments in plants and technologies.

To become a reality, the aggregation process had to overcome a number of hurdles and oppositions. In addition to a cultural problem, there were opponents among local politicians and suppliers unwilling to lose their influence on the municipal companies. Also, financial problems were to be solved: most of the utilities were not incorporated or joint stock but only public entities (*municipalizzate, aziende speciali, enti,* etc.). For those entities the legal rules—including governance, responsibilities, financing methodologies, etc.—were based on a totally different legislation that made it difficult to design and implement financing instruments like bond/shares issue or Mergers and Acquisitions.

[4]For a detailed discussion on the role of the Italian Authority, see Chap. 3 which is the contribution of Guido Bortoni, former President of the Authority.

The Observatory followed these dynamics from 2001 onwards.[5] Initially, the initiatives were mainly joint ventures or consortia (acquisitions were very rare), participated by two or more players, usually small or medium utilities sometimes with larger companies. Telecom companies also played a relevant role, with whom local utilities co-operated in the multi-utility strategy implementation. This point will be discussed later.

In the following years the aggregation process continued, reaching a peak in 2004, as shown in Fig. 1. This growth was due to the spreading awareness of the dimensional benefits and of the difficulties for the small and medium size companies to supply high quality service at reasonable costs. The law forcing the incorporation of the public utilities was also relevant, making the implementation of any financial deal easier.

Figure 1 shows the strategic focus of the agreements in the period 2000–2005. Horizontal integration—i.e. to perform the same business in a larger geographic area—is the dominating model; diversification was relevant in the first couple of years, together with vertical integration.

There are several notable examples of aggregation. The first significant case was that of Hera Group, born on the 1st of November 2002. Hera aggregated 11 local multi-utilities in the Emilia Romagna region. Since its establishment, Hera has had a balanced business portfolio: gas distribution, district heating, power distribution, water services, and other minor services. In 2002 the consolidated sales were €1.1 billion, with a net profit of about €36 million. The declared objectives of the aggregation were the following:

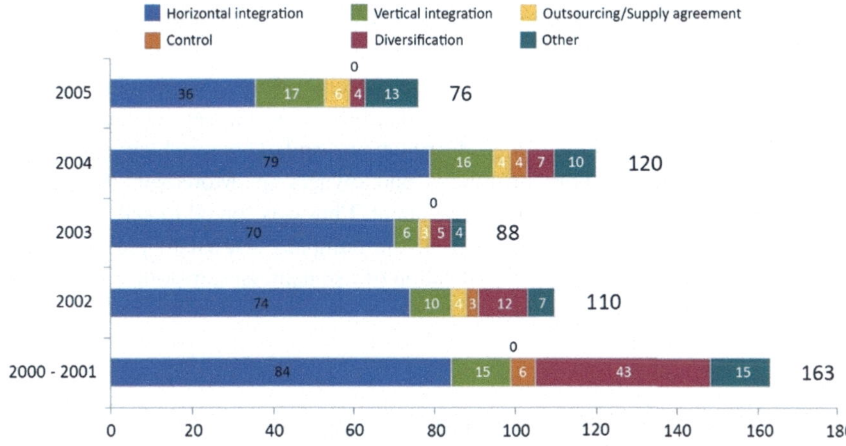

Fig. 1 Strategic focus of the agreement in the period 2000–2005. *Source* Agici (2006). Il consolidamento del mercato europeo delle utilities e le strategie competitive in Italia

[5]For more details see Part II: *Autori Vari, Le alleanze nelle utilities: tipologie e problematiche attuative*, The Observatory on Alliances and Strategies in the pan-European Utility Market, Milan, 2001.

(1) Rationalization of the portfolio, developing all the synergies coming from the aggregation
(2) Growth, exploiting the opportunities in the liberalizing utilities sectors and the strong territorial rooting.

Another noteworthy example is that of A2A, which started its activity in 2008 as the result of the merger among three municipal utilities[6]: AEM and AMSA from Milan and ASM from Brescia. AEM and ASM were active mainly in power generation and distribution and gas distribution, while AMSA was in charge of waste collection and treatment in the Milan metropolitan area. In 2008, the new-born A2A was the first local Utility in terms of revenues with €5.2 billion, coming from four areas: energy (power generation; electricity and gas distribution and sales); networks (power, gas and water); waste (collection, treatment and disposal) and district heating (cogeneration plants and networks).

A third case is Iren, born in 2010 as the result of an aggregation path which brought together five municipal utilities operating in Turin, Genova, Parma, Piacenza and Reggio Emilia. The process required several steps. In the first place, AMPS (Parma), TESA (Piacenza) and AGAC (Reggio Emilia), joined in 2005 to form ENIA. The following year, AEM (Torino) and AMGA (Genova) merged into IRIDE. These two new-born companies were active in two neighboring areas (IRIDE in the north-west of Italy across Piemonte and Liguria; ENIA in Emilia Romagna) and decided to merge in 2010, concluding a negotiation which initially included also Hera.

The only case that followed a different path—namely a disaggregation—was Enel. It was founded in 1962 with the objective to unify the Italian market for power; it became the incumbent in power generation, distribution and sale with some negative impact on efficiency. After more than 35 years, in 1999, the Bersani Decree established the unbundling of the electricity sector, thus forcing Enel to separate its generation, transmission and distribution businesses in three distinct companies: Enel produzione, Enel distribuzione and Terna, which should be independent in terms of both operations and ownership. Moreover, to spur an effective competition in the generation segment, Enel was forced to sell 15 GW of its capacity. Seeing its monopoly broken, the company started to pursue a new strategy, that led to a deep internationalization of its businesses, making it today one of the largest energy companies in the world.

5 The Rise and Fall of the Multi-utility Model

Among the strategies that many leading national and international companies tried to develop in the 1999–2002 period, the multi-utility model was one of the most relevant. The underlying idea was that managing various utilities' services together

---

[6]Companies which operated in the utilities sector, directly participated by the local Municipality.

could generate strong synergies both in infrastructure construction and operation, and in the commercial activities. In addition, many players included telecom services in the package provided to local users, based on the emerging broadband connections.[7]

An example is Enel, the mentioned Italian national champion vertically integrated in the power value chain, that in those years bought a telecom company, later named Wind, and carefully evaluated also the acquisition of the Acquedotto Pugliese, the second largest Italian water management entity. In addition, Enel expanded its gas distribution and sale business by taking over the largest private company in the sector, together with several other smaller businesses, becoming the second national player. Another example is Aem Milano (nowadays A2A) who extended its business to telecoms, exploiting the program to renovate the underground gas distribution network to deploy optic fiber; thanks to this project Milan become one of the first cities in the world in the ultrabroadband development (Fig. 2).

The largest European companies implemented this model mainly, if not only, through acquisitions, often in foreign markets. The consideration paid to the sellers was usually very high due to the strong competition among buyers. The latter saw a significant growth in their level of debt. With the summer 2002 stock exchange crisis the situation of many players became financially heavy; in addition, it became clear that the synergies expected from the integration of the different utilities sectors were lower than expected and also difficult to obtain. Furthermore, the managerial competencies necessary for example to manage the waste business were—and are—very different from the ones needed for power and gas, or water. Finally,

	Power	Gas	Water	Waste	TLC
Suez	✓	✓	✓	✓	✓
RWE	✓	✓	✓	✓	✓
E.ON	✓	✓	✓		✓
EDF/GDF	✓	✓	✓	✓	
Centrica	✓	✓			✓
Enel	✓	✓		✓	✓
Veolia	✓			✓	✓

Fig. 2 The multi-utility model for some European leaders. *Source* Agici (2004). Alleanze e strategie delle utilities: dalla dimensione locale a quella europea

[7]See the discussion on the multi-utility model in the 2004 Report of the Observatory titled: Alleanze e strategie delle utilities: dalla dimensione locale a quella Europea. The leading European companies invested in the triennium 2000–2002 around €70 billion.

all the sectors considered in the multi-utility model are relatively capital intensive, requiring considerable financial resources.

The results were several financial crises (for instance Vivendi, Suez) that triggered divestment processes, with the intention of focusing only on the sectors where the companies thought to be stronger. For example, Enel, with some difficulties, sold the telecom business; Germany's RWE liquidated the water companies in UK (Thames Water) and in USA (American Water Works). In any case, after these diversification experiences, the companies were different if compared to the beginning of the process: the Gas & Power bundled selling activity—the so called Dual-Fuel policy—is an example of a new integrated business policy.

6 The 2008–2009 Crisis. Competitive Reactions and New Business Models

The financial and economic crisis that exploded in August 2008 changed the utilities industry dramatically. If we consider the power sector, after half a century of virtually continuous growth, it suddenly had an unexpected halt from which the industry has never really recovered. As shown in Fig. 3, electricity consumption stopped to increase in 2008 and has stagnated ever since.[8] This abrupt slowdown in power demand was an unprecedented event, to which the utilities needed to adapt. In the next paragraph we discuss the impact of the crisis on natural gas.

At the same time, the unused capacity resulting from over-investments triggered a price competition that further reduced revenues and margins of virtually all the European players; Fig. 4 shows the price trends for households and industry in Italy, which follow different paths. Households price, as a result of the decrease in energy demand in 2008, experienced a sharp reduction in the years 2008–2011, and then started to recover slightly, even without reaching the pre-crises level. On the other hand, industry prices started to decline in 2012, when the recession and the sovereign-debt crises worsened in the Country.

Evidently, under these circumstances, generation costs became a key competitive variable: companies with a generation mix focused on hydro power or on other RES (with variable costs close to zero) were far better positioned than those based on fossil fuels. On the opposite, players with natural gas fleets were the ones who suffered the most: for several years, gas plants, although often brand new, were used for less than 30% of their capacity and were not able to repay the debts. In Italy, Sorgenia went close to bankruptcy and was bailed out by the banks that are now the controlling shareholders.[9]

---

[8]One should remind that nowadays electricity demand is decoupled from GDP growth; in other words, due to EE we are able to use less energy per unit of GDP.

[9]While drafting this chapter (December 2019) the controlling banks (especially Intesa Sanpaolo) are negotiating the sale of Sorgenia to new potential investors.

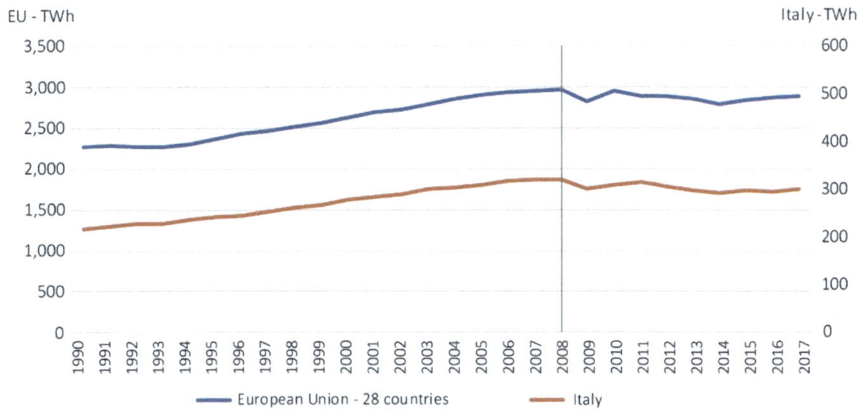

Fig. 3 Final consumption of electricity, 1990–2017. *Source* Eurostat 2019

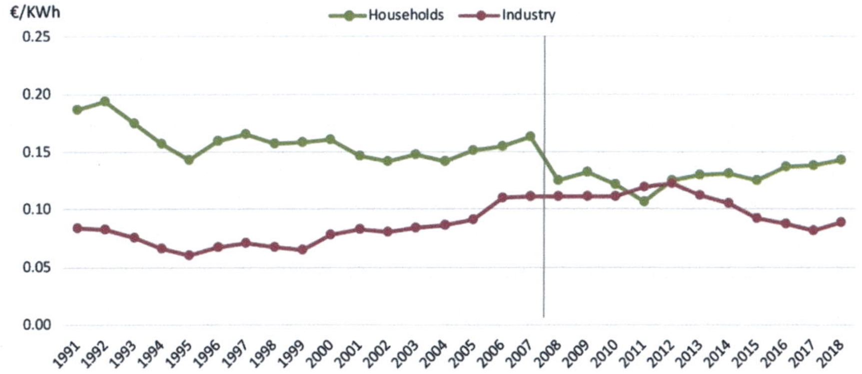

Fig. 4 Italian electricity prices for households and industries (€/KWh), 1991–2018. *Source* Eurostat 2019

Also, in Europe several companies did not immediately understand that a structural change was ongoing: they tended to consider the decline only as a cyclical event, so they expected the historical growth path to resume soon. In certain cases, they did not even stop their investments in generation plants, thus adding a factor leading to serious financial problems.

Which were the main countermeasures taken by companies to face the market and price decline? The most relevant are briefly indicated below:

(A) Cost cutting
(B) Freezing all the investments not strictly necessary
(C) Search of efficiency in the generation/distribution activity
(D) Selling the fossil-based plants, obviously at very low prices

Table 1 Additional services offered by various companies

	Distributed generation	Thermal res	Efficiency for electric systems	Efficiency for thermal systems	Energy saving devices	Insurance services
A2A		✓		✓		
Acea	✓	✓	✓	✓	✓	✓
Ascpopiave				✓		
AXPO	✓					
E.ON	✓			✓		
Edison	✓	✓	✓	✓		
Enel	✓	✓	✓	✓	✓	✓
ENI	✓		✓	✓	✓	✓
ERG						
GDF Suez (Engie)	✓	✓	✓	✓	✓	
Hera	✓	✓	✓	✓		
IREN	✓		✓	✓		
LGH				✓		
Shell						
Sorgenia	✓	✓	✓		✓	✓

Source Agici (2014)

(E) Developing the search for new revenues sources somehow related to the core business: i.e. offering extension.

This last point is particularly interesting. Table 1 shows the business extensions developed by various leading Italian and European companies[10] according to what they stated in their 2012 Annual Report.

The underlying idea was to extend the offering choosing products and services with a reasonable level of synergy with the traditional business and some attractiveness, in order to find new revenue sources. In most cases, one or more dedicated companies were created.

Although it can be considered contradictory (increasing efficiency means reducing power sales), many utilities in those years developed an offer in EE that included both services and products. Examples of services are systems to detect household power consumption or consulting on production processes efficiency optimization. Concerning the products, they include energy efficient air conditioners, LED lamps, boilers, and electric bicycles.

[10]The Figure is taken from the 2014 Annual Report of the Observatory entitled: Performance storiche e prospettiche delle utilities in Italia: Andamenti economico-finanziari e politiche di estensione dell'offerta, pag. 64.

7 The Importance of Natural Gas: From the Conversion to Methane to the 2050 Goals

Natural gas has played a fundamental role in the Italian energy system for decades; nowadays it is still very important, representing 35% of primary energy consumption: it is actually the Country's first energy source and it is reasonable to believe that this role will not decline in the next 30 years.[11] If we consider the natural gas consumption, it experienced a continuous growth from the 1940s to the 1980s as a result of the conversion to methane process. In the meantime, in Italy Exploration and Production started after the second world war and reached its peak in 1994 with a production of over of 19 bcm (Fig. 5).

In the late 1990s, natural gas had a sort of renaissance due to the liberalization process and the Kyoto Protocol Agreements to reduce CO_2. Gas consumption rapidly rose from 60 bcm in 1998 to 84 bcm in 2005: +40% with a CAGR of 4.9%. This trend was driven by power generation and energy intensive industries.

As far as power generation is concerned, the CCGT technology rapidly replaced many outdated thermal plants and became by far the leader for capacity addition due to its efficiency, flexibility and competitiveness. The share of natural gas on total power generation grew from 20% in 2000 to almost 60% in 2008. As for the energy intensive industry, natural gas replaced the other fossil fuels in manufacturing processes due to lower CO_2 emissions and higher efficiency. Natural gas consumption increased also in the residential and commercial sectors as old fired boilers were decommissioned (Fig. 6).

In the first years of the new millennium, the growth of natural gas consumption was expected to last for many years: the most important research centers and

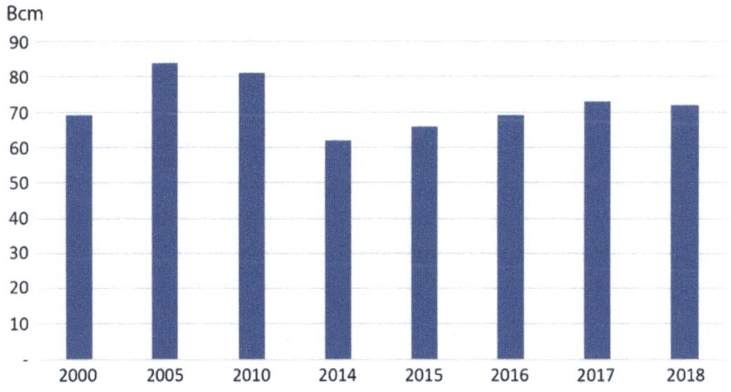

Fig. 5 Natural gas consumption in Italy from 2000 to 2018 in selected years, bcm. *Source* ENI, IEA, SNAM

[11]See the interesting contributions to this book written by Snam and Italgas.

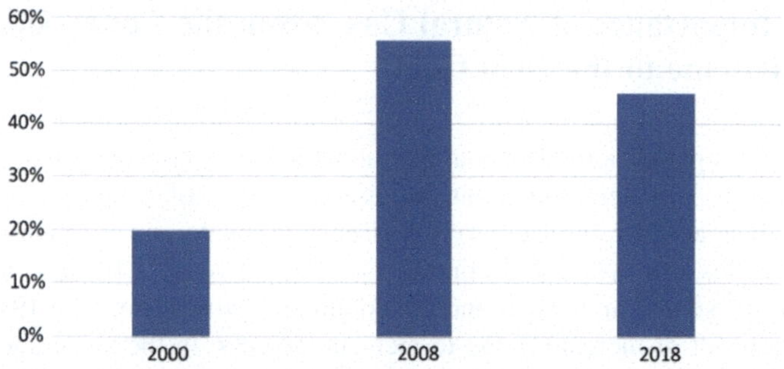

Fig. 6 Power generation from natural gas on total power production in Italy in selected years, %. *Source* AGICI on BP, Terna and SNAM

universities agreed that gas demand in Italy would reach 100 bcm by 2010. In the meantime, the national production of hydrocarbons continued its slight decrease and the dependence from imports constantly rose (in 2000 it was 76%, in 2005 it stood at 86%).

The strategies of the Governments and of the largest companies to face the rise of consumption and of foreign dependence were:

(1) The expansion of existing pipelines. The upgrades interested Transitgas, Transmed and TAG.
(2) The construction of new pipelines to diversify and secure supplies.
    a. In 2004, Italy was connected to Libya via Greenstream; the pipeline became operative in 2004 after only one year of construction.
    b. The TAP pipeline aims to connect Italy to Azerbaijan; the project started in 2003 and, after a lot of disputes and delays, it will supply Caspian gas to Italy in 2020.
(3) Development of LNG import facilities.
    a. The Adriatic LNG of Porto Viro was put in operation in 2009; it has an 8 bcm/year of import capacity and allowed Italy to receive LNG from Middle East, and in particular from Qatar.
    b. OLT Offshore LNG of Livorno became operative in 2013 after a series of disputes and delays. The terminal imports LNG from Qatar, Nigeria, Norway and USA.

Many other pipeline and LNG projects were suggested to the European Governments by the leading global gas players (among all BP, Shell, Gas Natural) as they consider Italy a sort of new gas Eldorado; most projects were not implemented due to different reasons. This turned out to be fortunate because, starting from 2008, the gas sector entered in a deep crisis as we saw for the power sector. There were two main reasons.

First, the economic crises that heavily impacted on gas consumption for at least half a decade. Second, the generous incentives introduced for solar PV plants, that made their capacity skyrocket from 483 MW in 2008 to over 18,000 MW in 2013. Solar production heavily reduced the CCGT plants utilization in daytime peak and in summer when electricity demand is high due to massive usage of air conditioning.

These two factors pushed down gas consumption; in 2014 it reached the threshold of 60 bcm, leading many analysts to believe in a progressive and unstoppable decline. Although it is true that the forecasts made in previous years have proved to be too optimistic, it is equally true that natural gas is regaining a central role in energy policies not only in Italy but also in Europe. This is due at least to the following reasons: (i) The gradual exit from the 2008 crisis was accompanied by a recovery in industrial production, which had a positive impact on gas consumption; (ii) The increasing weight of non-programmable RES made CCGT and gas peakers essential to stabilize the power system; (iii) The phasing-out of coal generation is creating some space for gas powered plants that have lower emissions.

If we consider the European 2030 climate goals and the proposed Italian National Energy and Climate Plan, both were perceived as a new menace to the gas industry. This is partially true. The Plan has ambitious targets in RES, EE, sustainable mobility as well as CO_2 reduction; this notwithstanding, the gas industry can develop in a number of markets that are eligible, turning again threats into opportunities. Below are some of the main examples.

- Sustainable mobility. The main Italian gas companies are investing in LNG to decarbonize heavy transportation (trucks, ships and diesel-fueled trains). Also, investments in gas-fueled cars are not negligible.
- Renewable industry. As mentioned, a big opportunity for the gas sector is biomethane from sludges, organic waste and agricultural scraps. According to CIB, the potential production can reach 10 bcm per year.
- Energy Efficiency and CO_2 emission reduction. Gas cogeneration, district heating and heat pumps can give a real contribution to the objectives. In addition, the Italian plan to decommission the coal power generation by 2025 considers the installation of new flexible gas plants for 5.4 GW.
- Hydrogen. For what concerns the most innovative project, SNAM is strongly focused on hydrogen production from clean and renewable sources.

To sum up, although natural gas consumption has never returned to pre-crisis levels, there are many reasons to believe that the gas industry is far from disappearing. On the contrary, it will play a leading role in the Energy Transition. In this process, the historical contrast between gas and power companies seems to vanish as the ambitious 2030 targets inevitably need the contributions of both industries.

8 Boom of the Renewables (RES). A Changing Generation Mix. Impact on the Energy Markets.

To make the situation more challenging, after 2010 we saw in Europe, and especially in Italy, a really explosive growth of the new RES.[12] In particular, solar capacity experienced a rapid rise after 2010, starting from zero and reaching 20 GW in 2018, with an extraordinary annual growth rate of 44.3%. Wind and bioenergy (biomass, biogas) increased at a slower rate, while hydro was already highly developed and remained stable. With this growth in installed capacity, RES now account for around 30% (33% in 2018 including hydro)[13] of the electricity produced in Italy (Fig. 7).

This boom was the result of incentives, often very generous, that attracted new players, namely investment funds, and in certain cases also insurance companies (for instance, France's Axa and Germany's Allianz), that became the main investors in the PV segment and, to a lesser extent, in the wind industry. After some years of major investments, several European countries realized that financial support was too high,[14] and nowadays subsidies are still available only in few cases; in addition, the auction systems used to assign the financial backing triggered a competition among investors reducing dramatically the level of public financial support.[15] For these reasons, we are witnessing an increasing number of players investing in grid parity perspective (or market parity), i.e. without subsidies.[16]

It is interesting to note that the European historical utilities made relatively low investments in RES; in most cases the effort was concentrated in onshore wind technologies. See Fig. 8 taken from a research done in 2012[17] which investigated the investments in RES planned by the leading European companies: onshore wind is by far the most relevant, followed by hydro (including pumping) and offshore wind. It is worth noting the low level of solar investments planned in those years by traditional utilities.[18]

---

[12]New RES were those generation technologies emerging in the period taken into account in this work. In particular, hydropower, although renewable, is not considered here because it has existed in significant dimensions since the last century.

[13]Terna, (2019) Dati storici.

[14]An estimate of the cost of RES incentives in Italy is in the range of €200 billion over a 20 years period; these subsidies are charged on the power bill and therefore paid by the final consumers. Other European countries, like Germany or Spain, made a relatively similar effort.

[15]The EU State Aid Guidelines required all EU member states to introduce by 2017 competitive mechanisms.

[16]Grid parity investments depend heavily on geographical characteristics and technological development that reduce LCOE.

[17]See the Report 2012 of the OIR Observatory: *Le strategie dei grandi gruppi europei nel mercato globale delle rinnovabili.*

[18]As we will discuss later, Enel is an exception being active in several RES technologies.

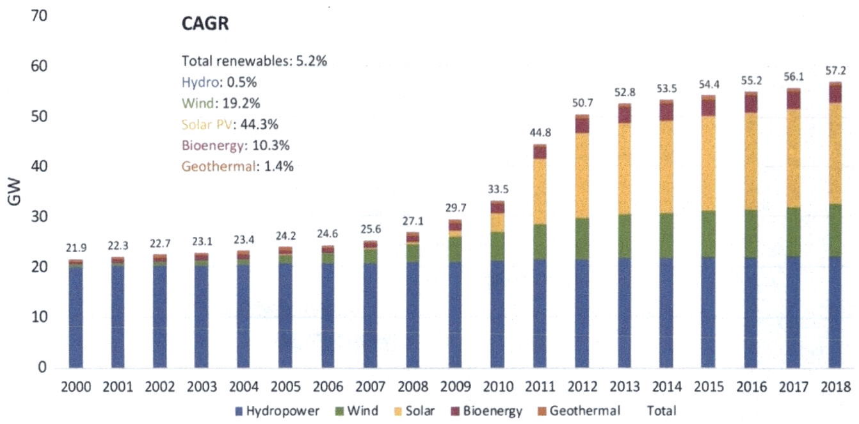

Fig. 7 RES installed capacity in Italy, 2000–2018. *Source* Agici on Irena 2019

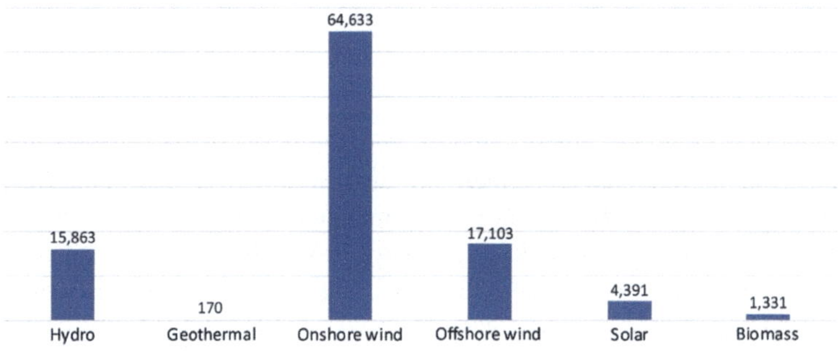

Fig. 8 Investments in RES planned by the 15 leading European operators (Acciona Energy, Alpiq, CEZ, E.ON Climate and RES, EDF Energie Nouvelles, EDP Renovaveis, EnBW, Enel Green Power, GDF Suez, Iberdrola, RWE, SSE Scottish and Southern Energy, Statkraft, Vattenfall, Verbund.) (MW). *Source* Agici on companies' data, 2012

This attitude was due to several reasons including a culture and competencies based on fossil fuels and on hydro power (the old RES) and the relatively high cost of RES. In addition, new RES could jeopardize the existing investments; besides, the new generation and distribution models based on RES were perceived as less reliable than the dominating model based on large production plants (with a capacity of several hundred MW), transportation networks and distribution grids. Last but not least, the peak-shaving effect has been relevant, i.e. the progressive elimination of the midday demand peaks due to the high performance of the new photovoltaic plants. Since prices used to have a related increase, this peak-shaving outcome cancelled a relevant source of profit for gas fueled plants.

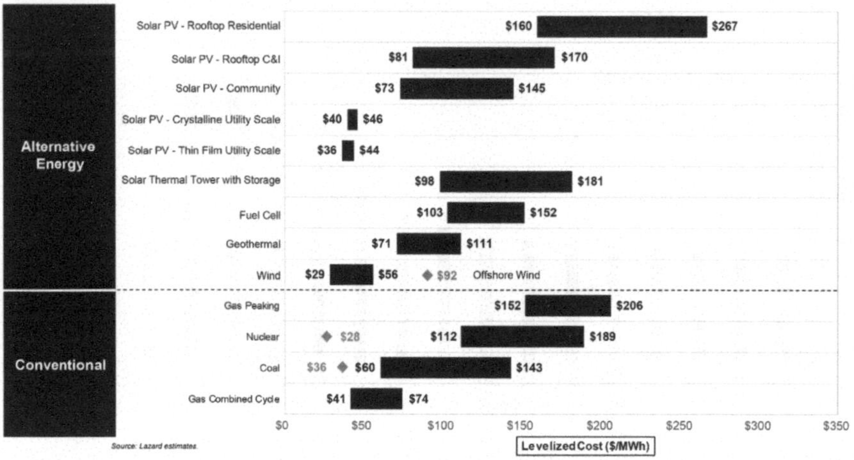

Fig. 9 LCOE of renewable and conventional sources (unsubsidised). *Source* Lazard (2018). Lazard's levelized cost of energy analysis

Currently, RES have a LCOE[19] which in many cases is competitive with fossil fuels generation (like gas, coke); this also because the variable costs for nearly all RES are close to zero (reducing therefore also the projects' risk) and the investments required less than two years to be implemented (Fig. 9).

The strong development of RES in many European countries had a relevant impact on the power market structure, creating problems that still need to be resolved. As already reported, gas generation was displaced due to the cost of fuel; dispatching priority put RES in a very strong competitive position; unpredictable production made it mandatory to create a back-up system to face sudden generation interruption. Transportation and distribution networks were (and are still now) to be rethought to fully exploit the new generation mix.

From the utilities' perspective, RES were not so appealing for the mentioned reasons. While the old RES were and are relevant, the new ones were virtually absent in the generation fleet of large multi-utilities like A2A, Hera, Iren and Acea. Only recently we witnessed a change in the attitude, but it is probably too late for them to reach a leadership or even a relevant position.

An Italian company that followed a totally different approach is Enel. The business line Enel Green Power (EGP) around 2010 adopted an aggressive strategy for a global development of its RES business (Fig. 10). While many European competitors focused mainly—if not only—on onshore wind, EGP developed its business on a wide spectrum, both technologically and geographically, with a

[19]LCOE is the acronym of Levelised Cost of Energy, that indicates under certain assumption the cost of power generation in a certain time horizon equivalent to the life of the plant. See for instance: *Lazard's Levelized Cost of Energy analysis*.

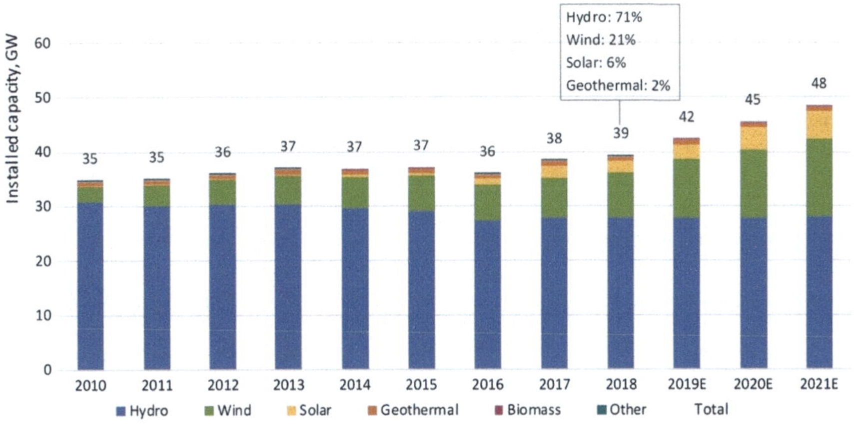

Fig. 10 EGP RES portfolio 2010–2021 *Source* Enel Green Power

portfolio that in 2019 is one of the largest and most diversified in the world.[20] In addition to hydropower, EGP is present in wind onshore, photovoltaic, geothermal and biomass. And this is true in the Americas, in Europe, and in Africa.

9 The Growing Relevance of the Environmental Challenge. Will European Directives Lead the Change Again?

In the last decade, environmental issues such as global warming and pollutant emissions, became central in the political and social debate, forcing policy makers to take concrete actions. This challenge became increasingly important in the EU policies: on 30 November 2016 the Commission published the *Clean energy for all Europeans* package (so-called *Winter Package*), a set of eight legislative proposals to facilitate the transition to a clean energy economy and to reform the structure and operation of the European Union's electricity market. The eight proposals (see Table 2) today have all been approved by the European Parliament and Council, and must now be transposed to national law in each Member State (MS). Table 2 shows the eight directives and their respective publication date and number.

[20]EGP's plants, divided by Country and Technology, can be found on the company's website: https://www.enelgreenpower.com/where-we-are.

Table 2 Directives and regulations included in the clean energy for all package

Directive	Official journal publication
Energy performance in buildings	19/06/2018—Directive (EU) 2018/844
Renewable energy	21/12/2018—Directive (EU) 2018/2001
EE	21/12/2018—Directive (EU) 2018/2002
Governance of the energy union	21/12/2018—Regulation (EU) 2018/1999
Electricity regulation	14/06/2019—Regulation (EU) 2019/943
Electricity directive	14/06/2019—Directive (EU) 2019/944
Risk preparedness	14/06/2019—Regulation (EU) 2019/941
ACER (Agency for the cooperation of energy regulators)	14/06/2019—Regulation (EU) 2019/942

Source European Commission

In a nutshell, the package revolves around a few fundamental pillars:

- Power Market Design. To design future power market based on more flexible and decentralized generation, on cross-border integration and power exchange, on increasingly active prosumers participating in the market thanks to smart metering, demand-side response, distributed generation and storage activities;
- RES. The RES coverage of 32% in the gross final consumption by 2030 is the crucial target for the Union as a whole, with MS required to individually cover this target in the range of 10–49% RES in the gross final consumption.[21] The directive also emphasizes the importance of prosumers by granting a right to sell their excess production to the grid without losing their rights as consumers;
- Energy efficiency. The proposal states a 32.5% EE binding target for 2030 determined for the whole Union, without defining individual national binding targets. In fact, it outlines that MS shall report to the commission the indicative national targets. They should be expressed as absolute levels of primary and final energy consumption in 2020 and contributions towards the Union's 2030 targets. As far as the energy performance of buildings is concerned, the Winter Package contains a revised directive, which defines renovation targets, addresses monitoring and control of energy use and tries to specify the regulations regarding the energy performance certificates.

With the Winter Package, the European institutions set a framework for energy related policies for the next decade, which will be gradually translated into new Laws by the individual governments in the next two years. This framework shapes

[21]It is relevant to underline that the RES directive did not define national binding targets for MS while it emphasizes the reduction of administrative barriers.

the context in which utilities are operating, pushing investments in the direction of the Energy Transition. In particular, the Directive on *Governance of the Energy Union* requires all MS to develop integrated National Energy and Climate Plans (NECPs) for the period 2021–2030 (and every subsequent ten-year period) based on a common template, that will translate the EU targets for decarbonisation into national binding targets.

10 Recycling and the Circular Economy. Myth and Reality

In the last 15 years the Circular Economy concept, made famous by the Mac Arthur Foundation,[22] has become quite popular. In truth, the idea of recovery and recycling that is an explicit pillar of Circular Economy was already well known and developed since the end of the 1800s especially in countries poor in raw materials such as Italy.[23]

The EU's commitment to a Circular Economy has moved to the forefront of policy discussions in recent years. Although the Circular Economy and resource efficiency have not been among the 10 priorities of the Juncker Commission (2014–2019), political and legislative activity has proceeded relentlessly on the topic, and important innovations have been introduced in the last decade, which have contributed to shaping the current market.

10.1 The Path Toward an EU Circular Economy Regulatory Framework

The current policy and regulatory framework on material efficiency and waste is largely rooted in the 2008 directive on waste.[24] The directive was the first effort to modernise waste regulation at the EU level, by introducing: (i) end-of-waste criteria specifying when waste ceases to be waste and becomes a secondary raw material; (ii) a waste management hierarchy providing for the prioritisation of waste prevention over re-use, recycling, (other) recovery, and disposal as a last resort; and (iii) the mandatory separate collection of waste by 2015 for at least paper, metal, plastic, and glass.

[22]See: www.ellenmacarthurfoundation.org. A Circular Economy is based on the principles of designing out waste and pollution, keeping products and materials in use, and regenerating natural systems.

[23]A comprehensive discussion of Circular Economy and of other related topics are in the Hera contribution in chapter.

[24]2008/98/EC https://eur-lex.europa.eu/legal-content/EN/TXT/?uri=celex%3A32008L0098.

The principle of subsidiarity, shaping most European policies, has been applied also in this field by providing for Member States, for instance, to take charge for the definition of end-of-waste criteria where these are not set at the EU level. The European Commission has been working, on the one hand, to lead the way with overarching regulations; on the other hand, catering to a significant diversity in the patterns of material flows in different EU countries, the Commission's approach entails empowering Member States, local and regional authorities, and the business community to be drivers of change.

The push for a more circular economic model was then boosted in 2011, with the issue of the communication "A resource-efficient Europe—Flagship initiative under the Europe 2020 Strategy",[25] and "Roadmap to a Resource Efficient Europe".[26] The following step was the publication, in December 2015, of the communication "Closing the loop—An EU action plan for the Circular Economy",[27] the first occasion in which the Circular Economy was specifically targeted from a policy and regulatory perspective. The action plan has been the basis for comprehensive reform enacted in the following three years. The core of the reform has been approved in 2018 with four landmark directives: the directive on waste management amending the 2008 framework regulation on waste,[28] the directive on packaging,[29] the directive on landfilling,[30] and that on batteries and waste electrical and electronic equipment (WEEE).[31]

The combined provisions of the four measures introduce significant innovations. Recycling targets have been updated by providing new targets for the separate collection of household waste, mandatory separate collection for construction and demolition (C&D) waste for a range of materials, a new mandatory extended producer responsibility scheme by 2025, a new food waste reduction goal of 50% by 2030, as well as new measures to halt marine litter. New regulation and targets are foreseen by 2024 for other topics such as non-hazardous industrial waste and reusable packaging, including measures for the diffusion of returnable packaging with deposit, the introduction of quantitative and qualitative limits to the use of single-use packaging, the introduction of economic incentive schemes. Among the most recent actions at EU level is the issue of the so-called single-use plastics directive,[32] prohibiting the manufacture of certain types of plastic products.

[25]COM (2011) 21 https://eur-lex.europa.eu/LexUriServ/LexUriServ.do?uri=COM:2011:0021:FIN:EN:PDF.

[26]COM (2011) 571 https://eur-lex.europa.eu/legal-content/EN/TXT/?uri=CELEX:52011DC0571.

[27]COM (2015) 614 https://eur-lex.europa.eu/legal-content/EN/TXT/?uri=CELEX:52015DC0614.

[28]Directive (EU) 2018/851 amending Directive 2008/98/EC on waste.

[29]Directive (EU) 2018/852 amending Directive 94/62/EC on packaging and packaging waste.

[30]Directive (EU) 2018/850 amending Directive 1999/31/EC on the landfill of waste.

[31]Directive (EU) 2018/849 amending Directives 2000/53/EC on end-of-life vehicles, 2006/66/EC on batteries and accumulators and waste batteries and accumulators, and 2012/19/EU on waste electrical and electronic equipment.

[32]Directive (EU) 2019/904 on the reduction of the impact of certain plastic products on the environment https://eur-lex.europa.eu/eli/dir/2019/904/oj.

10.2 Future Developments

The new Commission President von der Leyen has increased the Commission's focus on the Circular Economy which will be addressed as part of the framework of the "European Green Deal", the top priority as of von der Leyen's agenda.[33]

Following these ambitious policy orientations, it is likely that the legal requirements for all economic actors, and particularly for those concerned with the production, trade, transport, and consumption of material goods, will be impacted by ever more stringent requirements and objectives. The repercussions in the field of waste management are yet to be envisaged, however it is appropriate to expect that, as it happened to date with reference to European framework policies on energy and on waste, EU legislation will increasingly drive innovative changes in business models, new technology developments and applications, and organisational and process layouts. The envisaged direction will be increasingly oriented towards Circular Economic models, which will increase the competitive advantage of organisations already embracing the circular paradigm, and will be an additional reason for businesses that are late in adopting business models in line with the concept. As an example, following increasingly tight regulation on plastics, some economic sectors are already facing extinction or reconversion, such as single-use plastic cutlery producers. Similar challenges will impact other productive sectors as incentives and restrictions will make economic circular models more feasible and economically advantageous.

11 The Energy Transition. New Technologies and New Businesses for the Utilities

One of the main outcomes of the growing awareness on the environmental challenge is the Energy Transition: i.e. a long-term structural change in the electricity systems motivated mainly by environmental objectives and driven by new

[33]Part of the new Commission's commitment is to achieve the continent's climate neutrality by 2050. The main points set out for the deal are: increasing the European target of 2030 emission reduction from 40 to 50%; new international negotiations to increase the level of ambition of other major emission producers by 2021; a new Just Transition Fund to support the most affected by the transition, such as the more industrial, coal and energy-intensive regions; introducing a new biodiversity strategy for 2030, mainstreaming biodiversity priorities across policy areas; following a zero-pollution ambition to minimise air, water and noise pollution from transport, agriculture and food production, water quality, hazardous chemicals and other key areas; reducing the carbon footprint of the European transport sector and working towards the decarbonization of the maritime sector following a blue economy approach; promoting the Circular Economy and introducing the comprehensive "Farm to Fork" strategy for sustainable food, production to consumption; ensuring that tax policies deliver on climate ambitions, including reforming the Energy Taxation Directive and introducing new fiscal policies such as the "Carbon Border Tax".

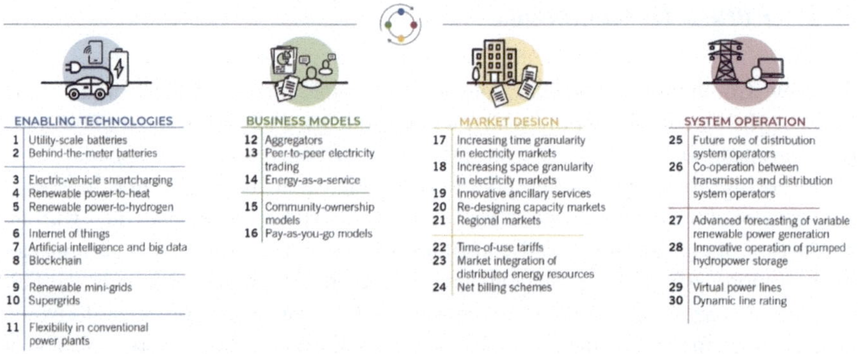

Fig. 11 Innovations involved in the deployment of RES in the energy transition. *Source* Irena (2019). Innovation landscape for a renewable-powered future

technologies, especially digital ones. The purpose was (and is) essentially to move from a model with large generation plants, using fossil fuels, to smaller ones, decentralized and based on RES. Another relevant objective is increasing EE both in power generation and in consumption.

This Energy Transition is anything but quick and easy: it is a complex and long-lasting radical change that involves a huge amount of finance, strong and public governance, technological developments, involvement of the population, and much more. RES are a key part of the transition. Figure 11 summarizes the four main innovation areas to allow the integration of RES.[34]

Those discussed below are among the reasons that make the transition to RES difficult to put in place:

(1) **Redesigning the energy production and distribution models**. This means to plan not only relevant investments but also the development of new technologies to fully implement the Transition. Digitalization is mandatory especially for the grids to make them smart; advanced grids allow for lower distribution costs, demand side management, better exploitation of distributed generation and of the prosumers, quicker response to accidents, and several other benefits.
(2) **Incumbents' opposition**. In many countries, incumbents have relatively old fleets based on fossil fuels, often because they have the related natural resources within the nation (for instance, coal or gas). In Europe, a good example is Poland where the high level of coal reserves does not stimulate the development of RES; besides, in many cases the national economy and employment are strongly linked to traditional sources and any change would generate tricky socio-economic problems.

[34]See Irena, Innovation Landscape for a Renewable Powered Future: Solutions to Integrate Variable RES, 2019.

(3) **Land occupation and the Nimby[35] syndrome**. RES are quite space intensive, requiring much more land per KW installed than traditional plants; in many European countries this is a growing problem while in the developing ones, where a lot of space at low cost is available, the topic is less relevant. The Nimby syndrome can be triggered also by environmental reasons as new RES spoil the landscape.
(4) **Authorization process**. Obtaining the authorization for plant construction is often complicated in many countries and could require a long time. This is particularly true in Italy due to the involvement of different layers of administration: State, Region and Local municipalities.
(5) **Stable electricity demand**. As the electricity consumption in Italy and Europe have been quite stable after the 2009 crisis (see Fig. 4), there is little need to add power capacity, so there will be new installations only to replace old plants which are being dismissed. Italy in particular has a generation fleet which is quite new and efficient, based mostly on gas-fired plants; so according to the opinion of certain observers, the available space for RES is limited.
(6) **Cost competitiveness of some RES**. We saw that LCOE of large-scale solar plants and onshore wind are now competitive with fossil sources, but for other RES this is not always true, such as Biomass and Offshore wind. This obviously makes investments in these RES technologies less appealing.

On the other side, RES have many advantages over fossil fuels. The main one is of course their (almost) perfect carbon neutrality. Given the effort that European Governments, including Italy's, are putting in the challenge to reduce emissions, a clean energy production mix is an essential pillar. This cannot be achieved without a large share of RES, especially given the fact that many countries (Italy, Germany, Austria, Denmark are just some examples) have already or are about to abandon nuclear power, which was the other big zero-emissions source of energy.

The utilities reaction to the Energy Transition challenge is multifaceted. According to the 2018 Report of the Observatory,[36] based on the analysis of leading European companies, several new businesses were under development, namely:

- **Smart grids and smart meters**. According to our analysis, intelligent grids enable several progresses; for instance, they allow better exploitation of the installed power capacity, particularly for intermittent RES; permit an active role of the prosumers; make it possible to react quickly to technological or natural problems/incidents. On the other side, smart meters allow utilities to extend the functionality of the services provided to customers, such as invoicing, energy management and monitoring of delivery conditions etc. Such intelligent equipment is complementary to smart grids in giving consumers the opportunity

[35]Nimby stands for: *Not In My Back Yard*. It refers to the fact that many infrastructure projects often find the opposition of the local communities, that will be negatively affected by the presence of the infrastructure.

[36]Andrea Gilardoni et al. Energy Transition. The European Leaders in action! Observatory on Alliances and Strategies in The Pan-European Utility Market, Milan, March 2018.

to participate in the market and the possibility of controlled individual generation and storage of energy. According to the EC,[37] nearly 200 million smart meters for electricity and 45 million for gas will be rolled out in the EU by 2020. This represents a potential investment of €45 billion for companies (utilities, IT and software companies).

- **Batteries and storage systems**. The diffusion of non-programmable RES generation makes it very important to develop solutions to face sudden power production drops, to stabilize the network and to supply power when the production from sun and wind is low or absent. Traditionally, the most popular was (and is) pumped hydro, but recently several different technologies were developed. For utilities, storage systems can be a good investment too: they can reduce constraints on the transmission and distribution network and thus can defer the need for major infrastructure investments. They can also become, if resold to the final customers or if used on the ancillary services market, a valuable income source. On the other hand, behind-the-meter applications allow consumers to manage their bills, reduce consumption of expensive electricity during peak demand and increase "self-consumption" from rooftop PV panels. Along with providing multiple services and user benefits, an electricity storage project can unlock multiple revenue streams from the provision of a range of services. Batteries are not yet a fully mature technology, but prices are constantly decreasing, stimulating the penetration of storage and changing the type of ownership of such installations. At the moment, most batteries are owned by the utilities, but in the long term the ownership will move towards final consumers. The falling price of batteries is also fundamental to enable the popularization of Electric Vehicles (EVs) and the spread of E-mobility.
- **Demand Side Management (DSM)**. Historically, interruptibility is a widespread practice around the world aimed at preventing malfunctions in distribution and generalized blackouts, appropriately governing the demand for electricity based on available supply. For a certain discount to the power price, energy-intensive companies (i.e. chemical, paper, steel, etc.) are available to stop production (and the use of power) immediately or within a certain notice. Today, thanks to the increasing dissemination of smart grids, meters and storage systems, the management of demand peaks relies on more innovative Demand Response (DR) tools and Ancillary Services (AS). DR systems consist of incentives for customers to lower or shift their electricity use at peak times, in order to help manage load on the network. AS support the transmission of electricity and the stability of the system providing short term corrections. Currently, most of the participants in DSM are not directly present mainly due to small size, but through Aggregators, i.e. companies that collect distributed generation/batteries and manage their capacity on the market. Utilities are

[37]EC (2019) Smart Grids and Smart Meters—website. Retrieved from: https://ec.europa.eu/energy/en/topics/markets-and-consumers/smart-grids-and-meters/overview.

beginning to enter this market mainly by acquiring Aggregators: a notable example are the two acquisitions made by Enel of *Demand Energy* and *EnerNOC*.
- **E-mobility**. In the last years we witnessed important developments in mobility electrification that involved not only cars but also bikes, mopeds, buses and trucks; currently the incidence of electrified vehicles on gasoline ones is negligible but there are widespread expectations of a relevant growth in the coming years.[38] From the utilities' perspective this trend could imply a positive inducement for a stagnant power demand and also generate some business opportunities; for instance, e-mobility development has a bottleneck in the diffusion of advanced and fast charging towers where the utilities can have a role in the construction of a capillary network. Also, the so-called Vehicle-to-Grid (V2G) technologies, that use the car batteries for balancing the power network, is a potential business.
- **Energy Efficiency**. EE is a potentially huge market that virtually cuts across all the economic sectors: industries, commerce, residential and even agriculture can improve their performance. Many technologies can increase efficiency and individuals' behavior has a major impact too. As we discussed, utilities were initially not so happy to develop a business that could jeopardize their core ones; nonetheless, financial incentives—together with the awareness that companies from other sectors could sell energy through the Trojan horse of efficiency—justified at some point the development of this business, which could be seen as an opportunity to increase the customer base, targeting specific types of customers (industry, public administration, but also households), and differentiate. Today utilities offer EE as a part of integrated services in view of extending the offer. An example is the energy management of buildings: utilities, thanks to their financial resources and knowledge, can take care of the entire useful life of an installation. To enter more deeply in this market, many Italian utilities have acquired Energy Service Companies (ESCOs), which offer EE as their core business.
- **Local and decentralized grids**. The RES development all over the world and the LCOE decrease boosted the market potential of local grids together with decentralized power generation. Both in developing and in developed countries there are many opportunities for designing and implementing local generation and distribution systems suitable for specific situations: from villages in Asian or African countries, to real estate developments, universities or military bases. Even the countless islands on all continents are interesting possibilities for local grids, becoming also possible alternatives to the expensive and environmental

[38]In October 2019 in Europe the electric fleet of passenger cars amounted to 580,000 vehicles, considering only full-electric ones (instead of hybrid). In Italy it amounted to 17,243. The National Integrated Plan for Energy and Climate foresees a growth up to 6 million electric vehicles, of which, around 4 million could be full-electric.

unfriendly diesel genset solution.[39] In Italy the potential for the development of local grids is not so high, and it is related to the provision of reliable and clean energy to remote areas and islands.

- **Ultrabroadband**. Although not key to companies' strategies, one should remind the convergence between optic fiber and the power industry, something already experienced when the multi-utility model was considered a winning one. Most of the companies divested from telecommunications. Only in few cases we saw some development based on the idea that the power grid could be used to deploy the optic fiber; in this direction, in Germany there is a cooperation between Telecom Germany and RWE for certain peripheral territories. In Italy, Enel in cooperation with Cassa Depositi e Prestiti and other companies, created a venture to develop this market: Open Fiber.

12 Development Paths in the Coming Decades

Today it is clear that utilities, across Europe (and probably worldwide), will operate in a context increasingly shaped by environmental concerns, mainly due to three factors.

First, the growing attention of citizens to climate issues (as witnessed by the success of movements such as Fridays for Future), that acting as customers, will require clean energy and will reward products coming from Circular Economy models.

Second, the policy framework: in the decade up to 2030, each EU Member State will have to implement its National Energy and Climate Plan, that will translate the targets into concrete measures for the decarbonization of the whole productive system and particularly of the energy sector. For the long run, the European Commission has presented its strategic vision, the so-called Long-Term Strategy 2050,[40] which aims at reaching a climate-neutral European economy by 2050. No doubt this is a very ambitious goal, but the Commission has identified seven actions crucial to reach it:

1. Maximizing the benefits of EE, especially focusing on "zero-emission buildings";
2. Maximizing the development of RES and the use of electricity to fully decarbonize energy supply;
3. Adopting clean, secure and connected mobility;

[39]"The global diesel genset market is projected to reach $17,821.3 million by 2024, registering a CAGR of 5.8% during the forecast period, according to P&S Intelligence". https://www.globenewswire.com/news-release/2019/05/29/1856758/0/en/Diesel-Genset-Market-Projected-to-Reach-17-821-3-Million-by-2024-P-S-Intelligence.html.

[40]28/11/2018—COM (2018) 773—*A Clean Planet for all*—A European strategic long-term vision for a prosperous, modern, competitive and climate neutral economy.

4. Fostering a competitive European industry and the Circular Economy as a key factor in reducing greenhouse gas emissions;
5. Developing adequate smart grid infrastructure and interconnections;
6. Achieving all the benefits of the bioeconomy and create essential carbon sinks[41];
7. Addressing remaining CO_2 emissions through Carbon Capture and Storage (CCS).

All of these, except for bioeconomy and CCS, were already identified in this chapter as new business for the future for utilities. It is straightforward that utilities have a great challenge ahead, but also a great chance to be leaders of the transition to a carbon neutral economic model.

Third, this effort called by the public and the policy framework seems likely to result in a wave of public investments across many Countries. Germany was the first to approve such a "Green New Deal" in September 2019, a package estimated to cost around €54 billion by 2023 which introduces a price of CO_2 (under the EU's Emission Trading Scheme) in transport and buildings, and simultaneously gives incentives for low-impact technologies. Italy too has moved the first step in this direction: the Update to the Economic and Financial Document[42] approved by the Government in September 2019, includes two funds at State and local level for a total €50 billion to be spent in 15 years. According to this Document, the resources will be allocated to urban regeneration projects, upgrading in EE and incentives to the deployment of RES.

To briefly conclude, along this chapter we described the deep changes that affected the utilities sector in the past twenty years, as a result of technological progress, market evolution and regulatory change. Nevertheless, utilities, both in Europe and in Italy, have retained their historical central role in our economies. In the upcoming decades, they will still be a key actor as long as they will face directly the challenges related to climate change mitigation and environment protection, through brave choices that will put them at the forefront of the "green" transition.

13 Structure of the Work

The book is divided into five main Parts consisting of contributions by a diverse set of actors that have been crucial in the transformation of the Italian utilities sector. First and foremost, the leading and most representative industrial groups that have been at the forefront of technological change and market dynamics in the past two decades. Besides them, the consulting firms that have advised the industrial players as they adjusted their strategy, operations, and internal organisation to the rapidly evolving market conditions. Then the financial institutions that have been

[41]This refers mainly to change in the land use, as reforestation, to create areas of vegetation that could absorb CO_2.

[42]Nadef, Nota di Aggiornamento al Def.

instrumental in fostering greater market concentration via mergers and acquisitions, as well as in raising financial resources for investment in major infrastructure construction and development projects. Finally, the regulator, whose policies have set the incentives for investment and growth in the sector to continue unabated, striking a balance between public and private interests.

In the next chapter, Accenture overviews the range of technologies that will accompany the switch from the older, centralised model of power generation and distribution to the emerging, decentralised one based on RES, prosumers, and microgrids. During this transition, the authors expect to see the two systems, old and new, coexist for quite a while in the frame of a hybrid model, even as profit flows and established roles in the sector undergo incessant change.

The chapter by Guido Bortoni, former President of Arera, the Italian Authority, focuses on the difficult task that the regulator has to fulfil in an industry landscape characterised by (i) deep, fast-paced transformation of the market structure and (ii) uncertainty regarding the technological trajectories, while at the same time being pressed to foster a steady flow of future-proof investment.

The chapter by Enel opens the second Part of the book, devoted to the industrial players that have developed a global reach. The result of the nationalisation of the Italian electrical industry in the early 1960s, Enel has gone from a traditional vertically integrated operator until the end of the 1990s to a DSO with a strong focus on RES and climate change, and an international presence spanning five continents.

In Chap. 5, Edison—a longstanding Italian company now controlled by French group EDF—emphasises its contribution to the development of both the Italian energy industry (electricity and gas) and GME, the Italian organised energy exchange. The paper stresses Edison's investment in RES, especially wind power, and the steps the company has taken towards decarbonisation, EE, the retail market and local communities.

In Chap. 6, ERG concludes the second Part by providing probably the most relevant example in Europe of business transformation. A family-owned enterprise grown up in the oil refining sector, in the 2000s ERG exited the oil business and shifted to energy with a clear-cut focus on RES. Currently, the group is a leader in Europe in wind power, with a number of projects to further expand its presence in other continents.

A2A (Chap. 7) inaugurates the third Part of the volume, devoted to large and medium-sized utilities with deep roots in the territories. Born of the merger between the municipal companies of Milan and Brescia in the economically strongest districts of Italy, A2A has developed a model of cooperation with smaller utilities whereby corporate tangible and intangible assets are shared locally within a frame that leverages and preserves local identities and knowledge.

In Chap. 8, Hera offers the portrait of a company mainly present in the Emilia Romagna and Veneto regions, again two of the most advanced areas in the Country. The paper revolves around the evolution of Hera's managerial approach with respect to social and environmental issues, highlighting what shifting from a CSR to a CSV perspective has meant for the group and providing examples of how the new vision has been implemented.

Located in the north-west territories of Italy (Turin and Genoa) plus parts of Emilia-Romagna (Piacenza and Parma), Iren (Chap. 9) has pursued growth mainly via M&A deals with utilities companies focused on different business lines. As a result, the Company has emerged with a rather diversified and balanced business portfolio comprising energy, water and waste management services.

CVA (Chap. 10) presents the interesting case of a local company entirely focussed on RES. Its portfolio, originally limited to hydropower—a reflection of its establishment in the heart of the Alps at the end of the nineteenth century—has further expanded since the early 2000s, with CVA starting energy production from both PV and wind plants.

Egea (Chap. 11)—yet another company with local origins—combines two distinctive features: a family-controlled ownership and the fact of serving a vast territory with numerous small towns and several large industrial companies. As the authors put it, being "smallest among the biggest" and "biggest among the smallest" enables Egea to keep abreast of technological and other national and international developments and to help local territories to deploy the most suitable solutions.

The fourth Part deals with national networks' operators. Terna (Chap. 12), the company in charge of managing the power transmission system, offers a glance into the challenges of planning and implementing infrastructure development. While planning is today highly determined by the Energy Transition objectives (e.g. integration of RES), construction and development have to address the most serious issue of social acceptance (i.e. the Nimby syndrome).

Chapter 13, by E-Distribuzione—the leading player (about 85% of the market) in power distribution—delves into the challenges that the development of RES and decentralised generation pose to DSOs. The picture that comes out is a detailed account of the technological solutions and R&D activities carried out be E-Distribuzione both in-house and in cooperation with university departments to fully support the Energy Transition and make the network clever, efficient and resilient.

Snam (Chap. 14), the Italian TSO for natural gas, has a key role in the Energy Transition, as this fuel is bound to play a major role for at least the next 20 years. The focus of the paper, however, is on the potential that hydrogen has to solve a number of pressing energy challenges in the medium-to-long term.

Italgas is the larger natural gas DSO in Italy. Its contribution (Chap. 15) focuses on remote-control systems and digitisation to optimise the management of the infrastructure and address issues of efficiency, safety, and sustainability. In the Company strategy, the integration between RES and natural gas is regarded as crucial to evolve an efficient electrical system. Biomethane and hydrogen both represent areas where the Company has a development potential.

Financial players take the floor in the fifth Part of the volume. Cassa Depositi e Prestiti (CDP) is a quasi-public body traditionally engaged in financing infrastructure and local administrations thanks to the revenues from postal saving. The paper (Chap. 16) illustrates the lending policies and financial instruments developed by CDP to support investment related to Energy Transition. Interestingly, the authors point to the challenges (and possible responses) arising for lenders in an environment where public incentives to RES development are being gradually phased out.

The last paper, certainly not the least, is by Intesa Sanpaolo (Chap. 17). The analysis here centres on the role of M&A transactions in reshaping the Italian market structure and the part the Bank has had in this evolution. Moreover, worth noting is the indication of what it means for a financial institution to embrace sustainability and how this may reflect in lending and financial policies.

Appendix

Publications of the *Observatory on Alliances and Strategies in the pan-European Utility Market*, that in 2020 has reached its 20th edition.

	Reports on the Italian market	**Reports on the European market**	**Working papers by Accenture**
2019	*Il ruolo degli operatori di rete e delle utilities italiane nella transizione energetica*	*Energy Transition: Role and strategies of European utilities and network operators*	*Flessibilità: Un'opportunità per la transizione energetica*
2018	*Cambiamento climatico e transizione energetica: Gli investimenti delle utilities italiane*	*Energy transition. The European leaders in action!*	*Oltre le smart city... «modelli collaborativi» per la valorizzazione del territorio*
2017	*Le strategie delle utility italiane di fronte alla sfida dell'innovazione*	*How to face the new revolution in the gas and power industry. Lessons from the 40 pan-European leaders*	*Decarbonizzare, decentrare, digitalizzare*
2016	*Elettricità, gas, idrico e rifiuti. Strategie e performance delle maggiori utilities italiane*	*How the 40 leading pan-European companies in gas and power and oil are facing in the difficult times*	*La digitalizzazione delle reti elettriche. Sfide e opportunità per il sistema elettrico nazionale*
2015	*Utility e competitività dei territori. Fattori abilitanti e strategie per un nuovo sviluppo*	*The new energy world strategies of the 42 gas and power leaders in Europe*	*The utility of the future: strategies to revamp utilities*

(continued)

(continued)

	Reports on the Italian market	Reports on the European market	Working papers by Accenture
2014	*Performance storiche e prospettiche delle utilities in Italia: Andamenti economico-finanziari e politiche di estensione dell'offerta*	*Winning moves to face a stagnant European gas and power market*	*Winning in the new energy marketplace*
2013	*I comparti delle utilities: Dalle ragioni della crisi alle strategie per il rilancio*	*Crisis of European energy market and strategies of the leading utilities*	*Digital power generation. Managing the key challenges facing European power generation in 2020*
2012	*Razionalizzazione e crescita delle utility italiane nelnuovo contesto nazionale e globale*	*Strategic actions in a time of sustained uncertainty*	*Developing winning moves for merger and acquisition integration in European utilities*
2011	*Performance economiche e piani strategici delle utility italiane tra crisi, privatizzazioni e sfide globali*	*A post-crisis outlook of the pan-European utility industry. Rationalization processes and new growth strategies*	*Post-M&A performance in the European utilities sector*
2010	*Il mercato italiano delle utilities. Verso il big bang delle ex municipalizzate?*	*The 42 leading companies of the European energy industry. Green economy, EE and energy independence: the role of pan-European utilities*	*A power shift: transforming the utilities offering*
2009	*Aggregazioni delle utilities italiane. Quali spazi residui?*	*The 41 leading companies of the European utility industry. utilities' strategy facing the financial and commodities shock*	*In search of the Optimal Fuel Mix (OFM) for the European union in 2020. Strategies and guidelines for European utilities*
2008	*Aggregazione delle utilities italiane. Quale assetto futuro del sistema Paese?*	*The 44 leading companies of the European utility industry. M&A market concentration and globalization*	*Pan-European gas industry scenarios. Truth and lies*

(continued)

(continued)

	Reports on the Italian market	Reports on the European market	Working papers by Accenture
2007	*Il mercato italiano delle utilities, tra internazionalizzazione e consolidamento regionale*	*Il consolidamento del mercato europeo delle utilities e le strategie competitive in Italia. Relazione generale*	*Proposte per una politica sull'energia*
2006	*Le prospettive del mercato dell'elettricità al 2010 in Italia*	*Il consolidamento del mercato europeo delle utilities e le strategie competitive in Italia*	
2005	(1) *Strategie delle utilities: sentieri per un nuovo sviluppo* (2) *Le strategie di marketing nel mercato liberalizzato del gas*	*Strategie e mercati utilities nell'Europa dell'est*	
2004	*Alleanze e strategie delle utilities: dalla dimensione locale a quella europea*	*N.A.*	*N.A.*
2003	*Le alleanze nelle utilities: strategie di integrazione e player emergenti*	*N.A.*	*N.A.*
2001	*Le alleanze nelle utilities: tipologie e problematiche attuative*	*N.A.*	*N.A.*

The Journey Towards the Energy Transition

Pierfederico Pelotti

Abstract This paper brings into focus the strategy issue. As the utilities sector undergoes sea change driven by technological progress, institutional reform and sustainability concerns, longstanding business models are rapidly becoming obsolete and established players are facing growing challenges to their role. In this effervescent, increasingly contended market environment, an extraordinary wealth of business opportunities opens up for companies of all sorts—from large energy firms to local utilities embarking on growth paths via alliances and aggregations, to startups and new entrants from side industries. After reviewing the key technologies impacting the sector and the main pillars of the EU energy policy, the paper delineates three macro-roles for future energy firms. It then stresses the function that smart grids, new digital capabilities, and market segmentation will have in redrawing the structure of the sector. Finally, it suggests that while a balanced mix of old and new technologies is likely to be around and even winning for some time to come, companies that resist innovation and change may soon find themselves as belonging to the past.

Keywords Business strategy · Business disruption · Digital capabilities · Market segmentation · Energy transition

1 The Journey Towards the Energy Transition

It is not a secret that the world is ready for the next energy revolution: decarbonised, digital, distributed. It is hard to predict who will win in the energy market in the next decade. There are currently many examples of different approaches:

- Large Utilities and Energy Companies are heavily investing in all key areas of the future energy transition, including many acquisitions and minor investments;
- Important local utilities play a role in the ecosystem with partnerships;
- Some traders want to position themselves as pure retailers with dedicated digital capabilities;
- Utilities with strong relationships with municipalities are looking to the next generation of smart cities;

A. Gilardoni (ed.), *The Italian Utilities Industry*,
https://doi.org/10.1007/978-3-030-37677-2_2

- Niche players with solid financials are designing eco-mobility;
- Some small new players are investing in the flexibility market ahead of other well established international players.

These are only few examples of how the utility and energy market is only at the beginning of the transition. There are many opportunities on the horizon for utilities of all sizes.

The 2030 target[1] is commonly considered the key milestone of the energy transition (as established in the "Clean Energy for all Europeans" package approved by the EU Commission). Some key actions include 30% renewable energy, and energy efficiency fully implemented in buildings and in major industrial plants. Part of the energy will be self-produced by energy communities integrated with their own micro grids. A contribution is expected even from the gas sector. An unpredictable drop in production costs could make technologies to produce and store hydrogen affordable. For example, it is estimated that the current price to produce and store hydrogen from excess renewable energy or from SMR (Steam Methane Reforming) will drop from $5/kg to $2/kg in 2030 and $1/kg in 2050.[2] And this scenario will introduce new options for the energy transition like producing energy with renewables and storing them with hydrogen, reducing the required peak of balance.

Utilities have an uncommon opportunity to lead the evolution and design a new power model with a new set of services. This will happen if they are able to quickly understand the transition and how to balance the commitment to the current business models with the incubation of new areas. It is clear that each utility or energy company can design its own value proposition based on its strengths, current position, ability to invest and the alignment to its core values.

A big threat is represented by disruptive offerings provided by new types of competitors working in lateral businesses close to utilities for mobility, energy efficiency, technology, trading, operations and maintenance (O&M) of plants. For example, Tesla is commonly recognised as car a manufacturer but they also offer energy services for residential with storage and photovoltaic modules. In addition, there is a dedicated brand (Powerpack[3]) with offerings for B2B storage and microgrids. At the same time, start-ups like Prosume[4] support the energy community's growth with a platform for peer to peer exchange and grid balancing. There are many other unconventional competitors in e-mobility, in trading, in engineering and construction beyond the meter offerings and in other new opportunities offered by the energy transition, described below.

In this complex environment, threats must be balanced with opportunities while identifying the right way to invest, collaborate and find profitable space in this new market.

---

[1]https://ec.europa.eu/energy/en/topics/energy-strategy-and-energy-union/clean-energy-all-europeans.

[2]European Hydrogen Manifesto—published by European Commission.

[3]https://www.tesla.com/it_IT/powerpack.

[4]https://prosume.io/: start up developing a platform for P2P energy exchange and grid balancing.

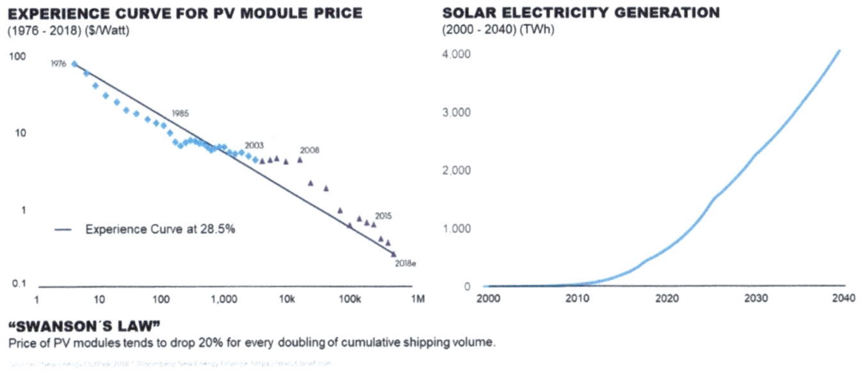

Fig. 1 Swanson's Law. *Source* "New Energy Outlook 2018", Bloomberg New Energy Finance

2 The New Normal: From a Centralised to a Decentralised Power Model

Significant changes are underway on the supply side as well. Technology advances and decreasing costs are making renewables more competitive. Some investment funds are selecting new renewables projects (like Octopus) that will be profitable without any incentive. By 2030, wind and solar could provide as much as 30% of the electrical power generated in the European Union. This energy will be cheaper than the electricity of traditional power plants, even if it will mean some important implications on network stability (Fig. 1).

Meanwhile, global uptake of behind-the-meter energy solutions such as rooftop solar by corporates is accelerating decentralisation. As a result of changes in supply and demand introduced by renewable energies, by 2030 utilities will need 30% more flexible capacity to balance supply and demand. Demand-side flexibility is key to grant the energy system efficiency, with energy supplied and consumed when it is needed. In this context, connected energy-using devices, vehicles, facilities, and industrial plants can provide demand response resources to the energy market by—for example—solving short-term imbalances in supply and demand. Furthermore, demand response can offer network value add, which is increasingly aimed at both transmission and distribution levels to avoid the costs of network development. In the future, digitisation could enable a host of additional services, allowing the involvement of demand response in the distribution of ancillary services.

Another major impact is the battery storage revolution, which is set to scale. Explosive maturation and growth of cheap battery storage is potentially the biggest energy disruptor for the future, driving a projected US$620 billion investment opportunity for energy storage (excluding pumped hydro) by 2040. Indeed, energy storage has the potential to dramatically change how we generate, deliver and use energy, by increasingly decoupling the timing of generation and utilisation of energy—unlocking a wealth of possibility for efficiency and arbitrage. Core to the

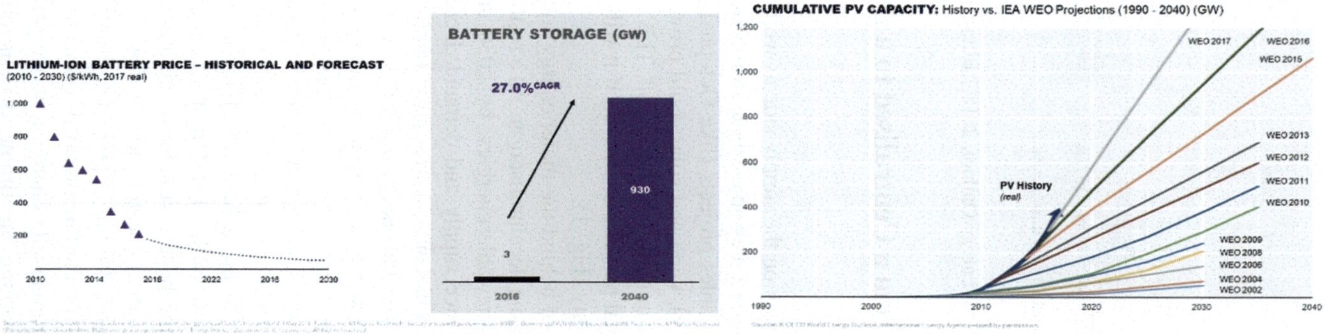

Fig. 2 Trends in battery prices and storage. *Sources* "Plummeting costs for wind, solar and batteries pose challenge to fossil fuels", Energy World, 1 May 2018, Factiva, Inc.; "Battery prices will go down as per BNEF", Commercial Vehicle, 13 December 2018, Factiva, Inc

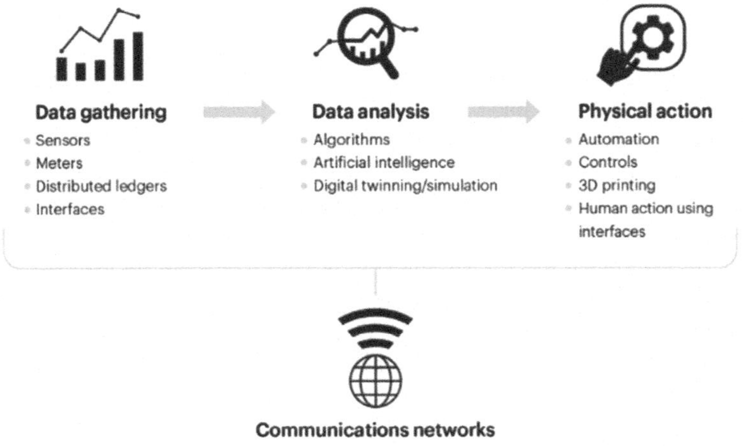

Fig. 3 Data driven value chain

new system will be multidirectional flows of energy and information enabled by widespread digitisation. Digitisation, through technologies that gather and analyse data, is used to make changes to the physical environment, which can help connect individual components that could achieve more end-use efficiency together than on their own (Fig. 2).

The following is a short description of key technologies enabling the energy transition that are used by utilities (but not only) to drive changes, automate, and generate new businesses as well as efficiencies.

Sensors—Sensors detect or measure input from the physical environment, such as daylight, temperature, motion or pressure. Even if sensors are not new, they have multiplied because of reductions in cost and size and advanced performance. Government policies have also favored sensor installation in the buildings sector, especially in the European and US markets, where building codes have guided the change over the last 15 years (Fig. 3).

Meters—Smart meters acquire high resolution information on real-time energy use, fails, reverse flow and other factors. They are now almost omnipresent, granting access to accurate analysis of energy demand efficiency opportunities. Furthermore, smart meters are simplifying bi-directional communication between residential or business meters and an energy retailer or a distribution network operator.

Distributed ledgers—Strictly speaking, digital ledgers are a way of storing data securely, rather than a technology for collecting data. However, because of their security features, distributed ledgers can support the data collection process by providing transparency and helping data providers and consumers have confidence that the data they provide, or use, is reliable and has not been tampered with.

Micro Services—Micro services are digital technologies that assist companies and individuals submit data to comply with policy requirements, such as mandatory reporting. This brings significant benefits for all parties, reducing reporting burdens for regulated parties while increasing the speed of collecting information and the quality of that information. In addition to their function as data gathering technologies, micro services are a key technology for converting data-driven analysis into real-world energy efficiency improvements, simplifying the interaction and supplying single self-contained functions.

Algorithms and artificial intelligence—Algorithms are "artificially intelligent" when they have the ability to "learn"—adjusting and optimizing their programming depending on data received over time, or self-correcting when their analysis seems unfeasible. Artificial Intelligence (AI) can draw meaningful insights from huge amounts of data, much faster than humanly possible, opening up new possibilities for identifying energy efficiency opportunities.

Simulation software and digital twins—Simulation software demonstrate how changes to an object or system would affect energy use. In the construction sector, Building Information Modelling (BIM) software can estimate the impact on energy demand from changes to a building's fabric, systems or occupancy. Advancements in BIM software have increased the reliability of computer-aided thermal analysis, reducing the performance gaps between expected and real energy performance in buildings. BIM has also underpinned more efficient building techniques, such as 3D-printed construction. Similarly, in industrial facilities and buildings, a "digital twin"—an exact digital replica of a physical asset within the production process—can be used to simulate and optimise how changes to its design affect its energy use.

Physical action and Actuators—Through machine-to-machine communications, or manually, digital data and analysis can be converted into a physical energy-efficient action automatically, via human actions in response to data and analysis. Actuators are devices that convert data into real-world energy efficiency actions. For example, a smart lamp contains a sensor that detects changes in lighting conditions. In industrial facilities, actuators are used in a wide array of applications, from opening valves in pipes to control the flow of gases and liquids, to moving heavy objects along a production line.

3D printers—3D printing is a good example of a technology that bridges the gap between data-driven analysis and the physical world to achieve real gains in energy efficiency. Every 3D-printed object begins as a software-generated digital model. The 3D printer itself (a group of mechanical actuators) uses these models as instructions for "printing" an object in 3D. Benefits include reducing O&M and supply chain costs and power plant downtime.

By offering both end-use and system efficiency benefits, digitisation is also reshaping views on energy efficiency and demand response: it is no longer possible to view the two processes as being separate, or in conflict. By joining up end-use energy efficiency with distributed flexible load, generation and storage, digitisation is helping to redefine the term "energy efficiency" to encompass both end-use and system efficiency. However, the utilities industry will likely operate for many years, even decades, in a hybrid model with both the traditional system and the new power model (Fig. 4: The New Power Model).

As utilities plan to pivot to the new power model, there is a huge opportunity for connected energy services—finding ways to deliver distributed generation and electric vehicle (EV) products and related services, energy management, energy efficiency and flexibility as part of consumers' new connected energy experience.

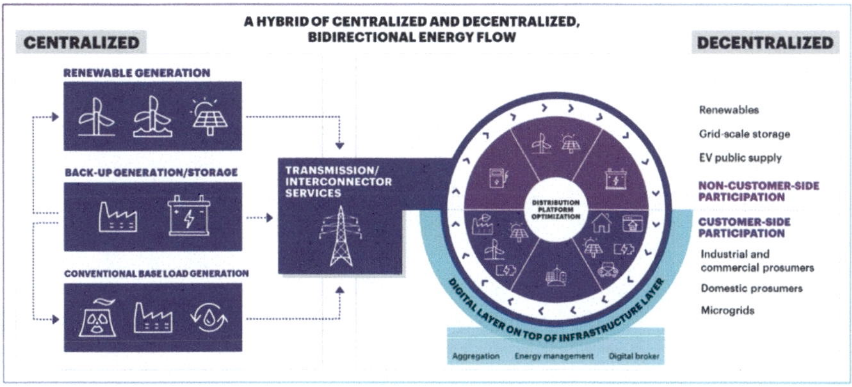

Fig. 4 The new power model

3 The Near Future

In Italy, the journey towards the transition is just starting to meet 2030 targets. The first step is to collect and summarise incoming regulations, targets, trends and finally create a holistic view to show how many opportunities will concretely emerge in the next ten years.

The EU Authority has approved a comprehensive update of its energy policy framework to facilitate the transition away from fossil fuels towards cleaner energy and to deliver on the EU's Paris Agreement[5] commitments for reducing greenhouse gas emissions. The completion of this new energy rulebook (see note 1 above) marks a significant step towards the implementation of the strategy, adopted in 2015. Based on Commission proposals published in November 2016, the "Clean Energy for all Europeans" package consists of eight legislative acts. EU countries have 1–2 years to transpose the new directives into national law (starting from the approval of a new policy in 2018).

The changes will bring considerable benefits from a consumer, environmental and economic perspective. It also underlines EU leadership in tackling global warming and provides an important contribution to the EU's long-term strategy of achieving carbon neutrality by 2050.

Key areas expected to improve in the journey to the energy transition:

- Energy performance in buildings
- Renewable Energy
- Energy Efficiency
- Electricity Market Design.

[5]https://ec.europa.eu/clima/policies/international/negotiations/paris_en.

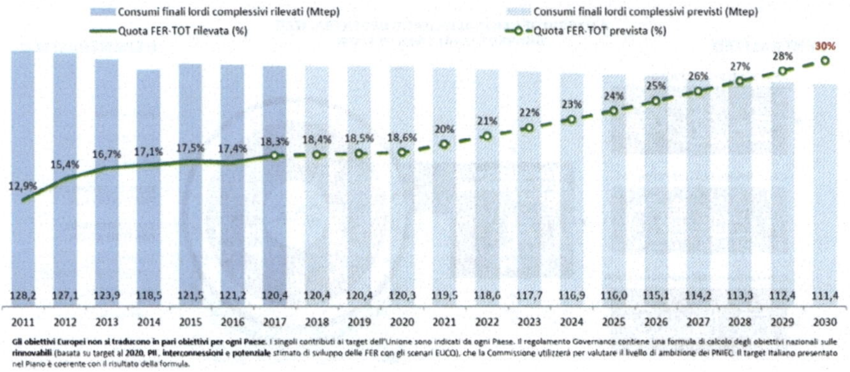

Fig. 5 The expected growth of renewable energy in Italy

The following is a short definition of these areas and the expected evolution:

- Buildings are responsible for approximately 40% of energy consumption and 36% of CO_2 emissions in the EU, making them the single largest energy consumer in Europe.
- the EU has set an ambitious, binding target of 32% for renewable energy sources in the EU's energy mix by 2030.
- Putting energy efficiency first is a key objective in the package, as energy savings are the easiest way of saving money for consumers and for reducing greenhouse gas emissions. The EU has therefore set binding targets of at least 32.5% energy efficiency by 2030, relative to a 'business as usual' scenario.
- A further part of the package seeks to establish a modern design for the EU electricity market, adapted to new market realities—more flexible, more market-oriented and better placed to integrate a greater share of renewables.

In Italy, a new set of rules will be released in order to meet the EU Objectives (PNIEC: Piano Nazionale Integrato per l'Energia ed il Clima). To meet the objectives, for example, renewable energy will need to grow by at least 1% each year. The Italian target is 30%.[6] This target is particularly challenging considering the energy demand affected by a new type of consumption redefined by electric vehicles: less predictable, more localised (Fig. 5).

The current FER contribution to transport is around 7% and is expected to grow to 22% in 2030, due mainly to electric vehicles (from cars to buses and from bikes to skates). Another big bet is energy efficiency. The current target seems to be very ambitious, challenging the previous scenario (2016 target), pushing the reduction of consumption to 43% compared with the reference scenario.[7] In the meantime,

[6]https://www.mise.gov.it/index.php/it/energia/energia-e-clima-2030.

[7]Primes 2007—baseline of the initial plan, consumption on 1990.

	Bassa Tensione Domestici	Bassa Tensione Altri Usi	Bassa Tensione Ill. Pubblica	Totale per mercato
LIBERO	11.821,5	3.685,2	214,4	15.721,1
	40,6%	*52,9%*	*80,7%*	*43,2%*
SALVAGUARDIA	0,2	65,0	26,5	91,8
	0,0%	*0,9%*	*10,0%*	*0,3%*
TUTELATO	17.293,4	3.220,6	24,6	20.538,6
	59,4%	*46,2%*	*9,3%*	*56,5%*
Totale complessivo	**29.115,1**	**6.970,9**	**265,5**	**36.351,5**

Elaborazioni ARERA su dati estratti dal SII

Fig. 6 Italian electric points of delivery by market

in July 2020, the liberalisation of the electricity market will be complete, impacting more than 20 million points of delivery as reported by the Italian authority for electricity and gas (Fig. 6).[8]

These objectives represent both a threat and a big opportunity for utilities. The threat is represented by the new set of capabilities required to address the energy transition and at the same time the effort for decommissioning traditional power plants while investing in new technologies. The traditional and protected business model based on large power plants with structured and efficient operations and maintenance is superseded by a decentralised power model with fragmented operations and maintenance. In addition, power demand in the next few years will probably be flat, supported by growing transport electrification. In this perfect storm, traditional businesses will progressively lose profitability and new businesses will require investments: new power plants, new technologies to manage the flexibility required by the new power model, more controls in real time on the grid, investment on EV (Electric Vehicles) infrastructure, pressure increase on the price of commodities.

Holistically, there are opportunities for utilities, allowing them to identify the specific mix and areas where they want to play a role (Fig. 7).

Services are organised in two categories: the role on the energy market and client segmentation. It is a simplification but it is useful to show a preliminary view of new offerings and business models for utilities in a single framework.

Simplified framework roles in the energy market:

- **Energy Provider**: the traditional role of utilities, proposing and managing energy and commodity offerings at a lower price and best service possible.
- **Connected Energy Service Provider**: the typical role managed in the past by energy service companies but in recent years increasingly linked with the ability to connect plants, data, devices, and clients. Most activities are performed with platforms able to manage a huge amount of data in real time and with the support of AI.

[8]Arera—Monitoraggio Mercati di vendita al dettaglio dell'energia elettrica e gas—2018.

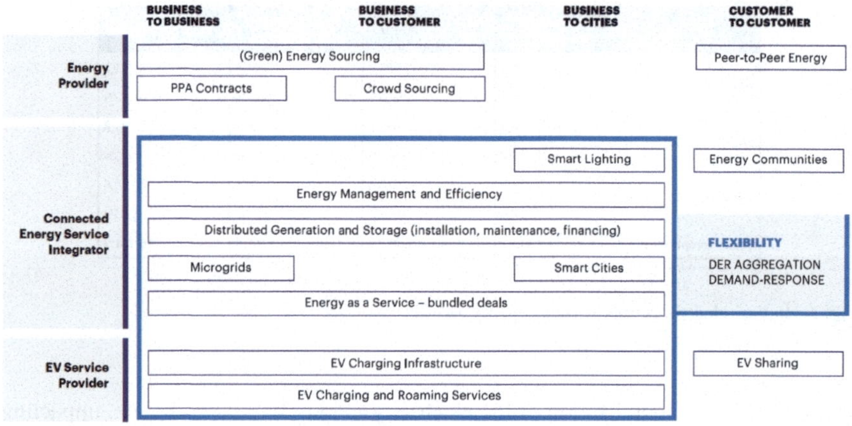

Fig. 7 Accenture energy transition framework

- **EV Service Provider**: this is a completely new market space that will grow with the electrification of transports. Services are not only connected with the energy and the technology to recharge vehicles (cars, buses, bikes, etc.) but they could be more focused on intelligent services, ecosystem management, operations and maintenance and customer service.

The proposed client segmentation is a high-level classification:

- B2B—Business to Business: services dedicated to enterprises, small, medium and large.
- B2C—Business to Customer: services dedicated to final customers, typically high volumes with low unitary margins.
- B2G—Business to Cities: services dedicated to cities from smart transports, to lightning and intelligent building services.
- C2C—Customer to Customer: services shared inside communities and peer to peer exchange.

The matrix describes a preliminary list of emerging services where utilities should play as a primary market maker. For example, the development of smart cities or establishing themselves as an aggregator for flexibility, or being the enabler of transport electrification, creating the infrastructure and a network for roaming among energy suppliers. All these services will be intelligent, proactive, digital, rapid and enable a set of other services used by other ecosystem players, like industrial partners for IoT, car manufacturers, telco providers or pure retailers, selling and managing the aggregation of services to consumers. To do that, all new services will be created on open platforms, modular, data driven and intelligent. Part of the income will come from unconventional sources like providing information, smart actions, controls, advice and consulting. For example, a car with embedded payment for energy recharging will find an available charging point and the recharge rate, or an energy community will request the best moment to release or acquire energy or will be managed as part of the grid. Finally, energy efficiency services will be mainly based on automation, forecasting, predictions and actions. In a word, we are moving away from a world that pays utilities for usage to one based on subscription services. Every company will have sustainability and circular economy objectives and will pay for that while reducing consumption of traditional energy.

In this future ecosystem, infrastructure will play the most critical role through the grid. Many opportunities are coming from a decarbonised and efficient energy environment, and "the grid" will play the hidden role of platform enabler. Surprisingly, the electric and gas grid could sooner or later be integrated. In the old energy model, big power plants convert mainly fossil fuels into electricity. The logistics of fossil fuels were complex but a deterministic problem, volume and efficiency were key to maximise profitability, efficiency, and reliability. The role of the networks (from fossil fuel supply until final consumption) was to ensure an efficient and reliable process, reducing failures and down time. Technical operations were a complex task for the scale of large power plants, the balance of supply and demand. The level of efficiency achieved was extremely high, in Italy one of highest in the world.

The energy transition will challenge everything that was previously achieved. Variables to operate networks are increasing, the same portion of network may be overloaded, under utilised, stressed for supply, for demand, for frequency. The level of inefficiency and failures should impact the energy transition. The complexity will require the ability to manage data, simulate and predict behaviours but this will not scare utilities that are planning investments in cybersecurity to create a more robust digital infrastructure, cloud transformation to be more scalable, digital implementing high speed communications, analytics and predictive modelling.

Accenture estimated an increase of 30–45% in extra flexibility from now until 2040 to enable the energy transition.

What seems to be more futuristic is the integration of gas pipelines and electricity networks. As mentioned, until now, points of contact of gas pipelines and electricity networks were power plants like Combined-Cycle Gas Turbine (CCGT) because of the conversion of fossil fuel energy into electricity. The dream of clean fuel may be a reality. The prediction of continued cost dropping for storage and photovoltaic panels may introduce a new dilemma. If anyone should produce energy with zero-marginal costs (no fuel required and limited O&M), the only production cost is the amortisation of the plant and to maximise profits required to continue to produce electricity and store it. Local integration between small and medium solar plants with storages are already available but with some constraints. What would happen to a mega photovoltaic power plant in the desert? In that case, the response is easy. Electricity transportation, for example, from Africa to Europe is quite inefficient.[9] In the past, there was a big initiative called Desertech to

[9] Energy Wasting for Joule effect.

produce electricity with solar plants in Africa (mainly desert areas) and transport the electricity to Europe via the largest direct current power line in the world to reduce inefficiencies (like SACOI for the connection of Sardinia with the Italian peninsula). The technology to produce blue (from the methane decarbonisation and CCS technology to store greenhouse gases) and green hydrogen (produced by electrolysis and stored in percentage in a methane pipeline) is accelerating its development. For the EU commission, the energy transition needs hydrogen's contribution. A good definition was recently made by Faith Birol, IEA director, "Hydrogen is today enjoying unprecedented momentum. The world should not miss this unique chance to make hydrogen an important part of our clean and secure energy future".

There are some tests on the Italian gas pipeline to confirm that introducing 10–15% of hydrogen in transported gas is secure and risk free. In that case, hydrogen storage in the gas pipeline is an efficient way to transport electricity generated remotely if the costs of hydrogen production will significantly drop from \$5/kg to \$1/kg. The use of hydrogen could change many other things, for example balancing the network: using fuel cells in congestion nodes instead of other types of balancing. This scenario will be complicated if hydrogen storage competes with other flexibility markets or virtual power plants. In any case, utilities are not alone in this journey and need to pay attention to the competition. Some new competitors are trying to change the game and create new "blue ocean offerings" where it could be difficult for utilities to compete.

4 The New Pool: A Complicated Game That's Not for Everyone

The opportunities described in the previous chapter are attracting new entrants in the energy market along the entire value chain. It is true that in the past ten years most utilities underperformed total returns for shareholders compared to their respective industrial stock index and it is true that many utilities faced issues with the decommissioning of nuclear and coal power plants. At the same time, it is true that the energy transition will change the game, some traditional barriers (invested capital, government agreement, vertical integrated value chains) are falling and adjacent operators are considering entering into new services.

The outlook seems to be very challenging because there are typical ingredients for what we call "business disruption". Business disruption is when an established business may be destroyed by trends attracting competitors able to offer similar services introducing new capabilities (typically digital or business platform based). The energy transition may bring business disruption if utilities won't investigate the forward game.

As discussed, digital capabilities will play an unprecedent role and in many cases innovative capabilities should attract expert players in such fields. Many energy transition businesses will be data businesses and companies like start-ups in AI or

Internet of Things (IoT) could be more reliable, quicker and innovative than old utilities. In addition, other energy businesses will be "asset based" businesses like e-mobility and manufacturers (automotive or industrial). They may think of being more effective in building or designing their own charging stations. Another example is related to all businesses with services like smart cities or energy communities. These topics are becoming very popular from students to politics, from start-ups to digital giants. Everyone in these categories feels like they might be the one to change (or save) the planet in terms of efficiency, sustainability, ability to execute and finally, ability to make profits. And this pressure should introduce new variables. Some competitors may decide to enter these businesses without making a profit and instead focus on experiences. This will create additional pressure on the market and especially on utilities.

In such an environment, it is hard to predict who will win and why but it is commonly accepted that understanding and managing the ecosystem will play a critical role in navigating and finding positioning. In the last chapter, a possible framework to navigate these threats and how to balance risks and opportunities will be described.

5 Who Will Be Your Next Competitor?

Conventional utilities are under pressure driven by decreasing demand, load factors, commodity prices and deregulation—moving to a new power model—with consumers taking a more active role. Addressable power demand in developed markets could drop ~20% or be flat in a more optimistic scenario over the next decade. Profit growth will be dependent on renewables, new products and services, and operational transformation. In the future, utilities will explore new businesses and look to expand their product offerings with data as the pillar of innovation. Each utility will strive to differentiate itself and aim to be unique from its peers, relying on the new opportunities technology and the digital economy offer. Organisations are shifting away from the centralised power generation model which means that utilities will increasingly operate assets that are outside their asset portfolio, which they do not necessarily own. Digital is transforming the 'how' and 'what' services being delivered to business leaders in utilities. New capabilities are enabling a cross-functional, collaborative workforce capable of delivering greater insights.

The industry landscape is changing, new profitability and growth targets address declining margins, driven down further by new market entrants and regulatory interventions. Oil companies are now joining the utility market through acquisitions and have aggressive growth plans (for example, Shell or ENI). Additionally, rapid advances in technology such as cloud, automation, blockchain, big data and analytics are driving opportunities that have immense potential for utilities.

We can't forget that there is an increasing need to provide an omni-channel experience that allows consumers to contact utilities when it is convenient for them. Leveraging digital contact strategies to improve customer satisfaction and offering

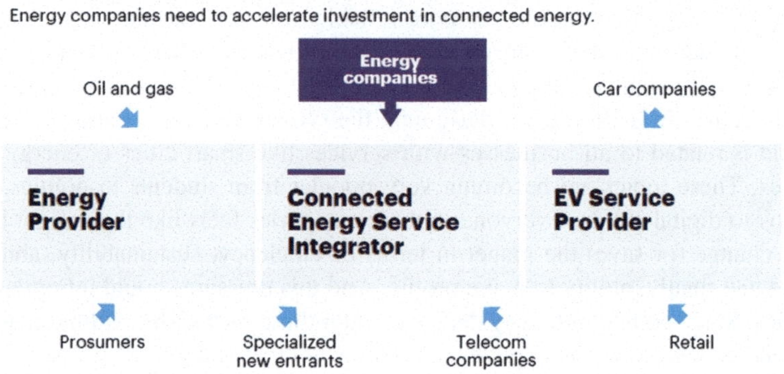

Fig. 8 New utilities competitors

new products and services are needed in order to retain existing customers and to grow new revenue. Speed to market is imperative to ensure a competitive market advantage.

The value pools are significant. The opportunities are compelling. As a new ecosystem develops, new entrants are investing in connected energy solutions for future growth (see Figs. 4 and 8).

In a world where oil prices are declining, the oil and gas industry is looking for new growth strategies, with electricity emerging as a key component. Some global oil and gas companies are leveraging mergers and acquisitions to accelerate transformation by investing in renewables and energy management-related services and by aggressively acquiring start-ups. For car manufacturers, the focus is no longer on solely selling EVs. Major players are now looking to combine EVs with smart, data-driven energy solutions. Some are even creating new companies to allow people to produce and sell renewable energy, with or without an EV, as well as offering storage solutions that could be connected to energy management systems. Meanwhile, we see the rise of a new generation of start-up companies with provocative value propositions that span flexibility services, peer-to-peer (P2P) energy trading, connected home solutions, and e-Mobility services. And digital giants are taking bold positions in areas like smart homes and energy management services (Fig. 9).

6 Golden Rules to Reshape the Next Operating Model

Many things will change on the energy market at an unprecedent pace, profits will flow into a new value chain and many established roles may change. When innovation and disruption is coming, it is not easy to act quickly.

The risk for utilities is to underestimate energy transition trends while maintaining the current operating model because it is creating enough profits. The concept of a

Fig. 9 Competitors by value pool

wise strategy[10] is mixing actions in new businesses, while investing in the efficiency and reliability of legacy businesses. The core business is key to funding and growing "the new".

The strategy may be organised in three main areas:

- Grow the Core: investment supporting the expansion of legacy businesses, like retail or grid modernisation
- Transform the Core: take benefits from new technology and digitalisation to reduce costs, introduce machine learning and AI in repetitive and low value tasks (e.g. AI supporting customer or grid operations)
- Scale the New: incubate and foster new businesses, identify profitable business models in the energy transition space (e-mobility, energy efficiency).

The energy trend is clear, policies are in progress and it is important to identify, test and scale new businesses. The first step for utilities is the key areas of current legacy businesses. In Fig. 10 (Operating Model Framework to success in the Energy Transition), preliminary areas of core businesses where utilities could struggle were listed.

Looking to future businesses, the situation is more complex. The investments and capabilities required may be reasonably uncovered and far from the current business models.

There are different approaches to organise the skills needed to enter and navigate new businesses: acquiring skills, building skills, and enabling skills.

- The approach of "acquiring" is typically implemented when an established utility acquires start up ecosystems or selects niche players to work with to develop new offerings or markets. This approach requires cash availability and/

[10]Pivot to the Future—Omar Abbosh, Paul Nunes, Larry Downes.

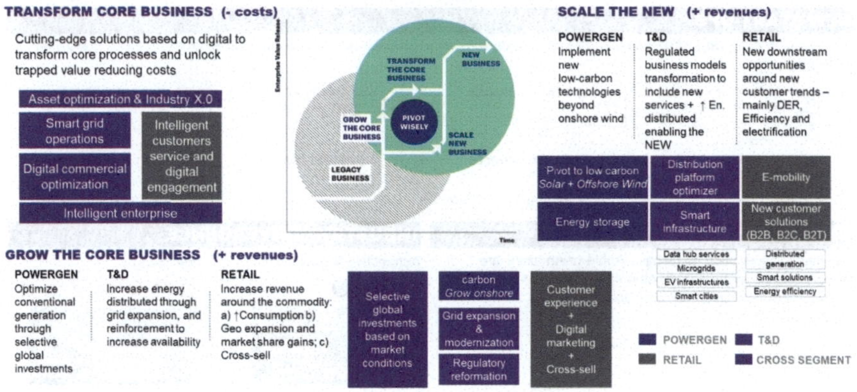

Fig. 10 Operating model framework to navigate the energy transition

or bank/investor support. It is reported that less than 50% of these acquisitions fail and it is a critical skill to maximise the return on such investments. This approach resolves some issues like creating a culture for new businesses, attracting digital native talents and having the opportunity to foster a new company purpose different to the current company purpose. At the same time, the "acquiring approach" generates risks of spending money in the early stages of innovation when profits are low and start up evaluations may be high compared to the real return.

- The second approach is "building". As described, the assumption "new businesses will be digital businesses" is at the core of part of the energy transition. Thus, some utilities believe they can transform and be recognised as digital service providers able to produce and commercialise software. Market analysts agree that the value will be on services rather than commodities. Following this assumption, some utilities are working on being digital and software providers for energy transition businesses like e-mobility, energy efficiency, micro grids settlement, real time network balancing, aggregators, etc. The idea of becoming a software factory is an interesting perspective because of the combination of energy and digital capabilities. The complexity of this shift is to be able to maintain the solutions and services in line with software and digital market specialists. The competition in such fields is dominated by digital giants able to make big investments. There is the opportunity in some niche energy services that are a low priority for digital giants and energy knowledge will be more important than the ability to invest money.
- "Enabling" is the approach that every investor wants: Low investments, strengthening the current position on the core energy market, gaining from others without limits. Such an approach is commonly defined "business as platform" and typically works by connecting consumers with the service and making a profit from the connection of sparse suppliers. The platform is the enabler for suppliers to sell their services. Companies like Uber, Amazon,

AirB&B are working with this model. Applying this model to utilities means establishing that the main value delivered is the services offered to consumers enabling heating, light, cooling, transportation, efficiency, security, etc. In this model, the complexity is to manage consumer needs in an extraordinary way while connecting the ecosystem not as a supplier, but as part of the game.

After each utility has selected its approach and identified key services for the future, a critical point is to bring results as fast as possible to benefit from uncertainty and enter new spaces before "the new" becomes "commodity". Typically, converting strategy into results is a complex task. Three golden rules can help; these have been tested in many other similar markets with the same ingredients.

Rule 1: Agility. Agility means making decisions, evaluating risks, launching projects and always being customer-centric, with the ambition of testing deliverables. Embracing agile methodologies should help organisations grow quickly with less investments, while selecting and attracting new talents, typically young, and motivated to impact and bring purpose to their tasks. Some agile organisations may fail fast but acquire skills for future success with an adequate retrospective on the output.

Rule 2: Disrupt your business. When the trajectory of future business developments is clear, protecting legacy businesses that are slowly declining may distract from the real priorities. Typically, the fear of change is the enemy of future business growth. The future requires investments, talents, vision, ambitions and courage. Not everyone has these qualities. Achieving new objectives while maintaining the same rules isn't possible. Some utilities who achieved this, did it with a very aggressive footprint, shifting power generation only to renewables or investing only on new value propositions while managing operations on the "core".

Rule 3: Innovate with Technology. Innovation is at the core of many operating models of high growth companies. Technology doesn't necessarily mean innovation but innovating with technology is the way to leverage technology to innovate the business.

Innovation must be placed at the centre of the organisation: the ambition to be recognised as an innovator and foster an internal culture of change, to evolve and invent new value propositions for consumers, for operations, for corporate processes and in all key areas.

It is proven that organisations who move towards innovation are more profitable, more open to change and to collaborate with the ecosystem.

To conclude, the energy transition is coming and it is ready to disrupt legacy energy businesses while creating new areas and opportunities. Utilities are well positioned on the market, less protected than in the past from regulations but still protected by their knowledge of energy trends and mechanics. The transition will accelerate in the near future under the requirements and expectations of European targets to meet the Paris Agreements to reduce the impact of climate change. The competition may increase in energy services and in particular in those areas that could sustain future growth like e-mobility, energy efficiency and energy communities. Utilities have the challenge to rotate their priorities in this environment, reframe their priorities, strategy and approach to execution. To improve execution performance, some golden rules are suggested.

For all utilities that are still waiting to embrace the revolution, no one knows how fast the transition will go but everyone knows the world of energy will change. All utilities should be part of the revolution and those who remain on the outside, will probably soon discover they are part of the past.

Regulating by Uncertainties the Energy Sector in the Deepest (Ever) Transformation, While Fostering Future-Proof Investments

Guido Bortoni

Abstract This paper brings into the picture the perspective of the energy regulator. It casts light on the challenges associatsed with the task of guaranteeing a number of medium-to-long-term objectives such as an adequate flow of investment in infrastructure, development of well-functioning markets, and a balance between private profit and the general interest. In particular, the paper addresses the concern of fostering long-lasting, "future-proof" investment in a sector that has been characterised by intense technological change and fast-changing market conditions—both sources of major uncertainty as to the effects that regulatory decisions will have on future developments within the sector. Drawing on the author's personal former experience as president of the Italian regulatory agency from 2011 to 2018, the paper illustrates and assesses past regulatory action with reference to (i) reform of the gas market, (ii) adoption of a new generation of smart meters, (iii) reform of the electricity retail market, (iv) and integration of RES into the electricity market. The review illuminates the multifaceted circumstances that have fostered—or hindered—successful forward-looking regulation.

Keywords Regulation · Uncertainty · General interest · Energy transition · Gas market · Electricity market

1 Bit of Today's Regulatory Dilemma

Starting with a common play: "Easy to say, hard to do".

It could be argued that this common play applies to a large part of the human activities that are planned to take place in the future and the energy sector does not make any exception in this regard. Indeed, planning energy commits ample time periods in the future and, usually, the results of those actions are in no way immediately consequential.

Furthermore, this common play becomes a real dilemma that also strikes in a dramatic way if one or more of the following circumstances apply: (a) the agent of

A. Gilardoni (ed.), *The Italian Utilities Industry*,
https://doi.org/10.1007/978-3-030-37677-2_3

the 'said but not realised' is an energy regulator; (b) the matter has to do with investments in markets and infrastructures in the same sector; and (c) the time in which we live is, like today, characterised by an intense and profound energy transformation affecting everyone: institutions, citizens and businesses.

Let us see why the situation is tinged with drama for each of the three circumstances indicated above. It is always recognised that the entity issuing the sectoral rules—say a regulator—should act as the first stabiliser in the system, a body giving certainty to investors and who is met by trust in conducting this role. If the mechanism functions virtuously, the outcome is very positive: risk profiles reduce, the economics of investment reckon with expectations of reasonable return, and prices and tariffs are lowered for consumers. However, the regulator is not a legal person liable to bear the risks of the system, nor is it easy to compensate for any negative results of such risks other than weighing up the user community. In many cases, even in sectors other than energy, users are found to bear the costs of errors or inadequate regulator decisions. An example in energy is the unhappy disclosure of *stranded costs* borne by consumers as a result of investment decisions in a technology or business model that fail to find remuneration in the market or go too quickly out of the market. Therefore, the conscientious regulator has always to pay attention to not just not making mistakes (although it is inevitable for anyone) but to adopt the "most informed" decisions.

The other two circumstances are used to inform the regulator as far as possible.

Investment concerns complex, technological and behavioural objects and its effect on the sector must be fully understood in both the short and the long term to enable the regulator to design appropriate instruments to promote investment that also serves the overall purpose of the system. A choice of investment in energy, even if carried out by market operators, must respond to economic signals that, in whole or in part, bring the investor's interest in making a profit in line with the general interest of the system. Only put in this way, market action is helpful in pursuing general interests and at the same time essential to achieve maximum efficiency. Market signals and instruments in the regulator's toolbox are properly designed only on the basis of a thorough knowledge of the sector, its context and its dynamics. A complex phenomenology must also be deeply understood insofar as the market can experience failures, intrinsic to its nature, that would cause failure to achieve positive externalities, despite the fact that such externalities would be beneficial for the system and thus worthy of being reached. In such cases, a pragmatic regulator anticipates the potential failures and devises remedies to overcome them by putting the market under the (right) conditions to pursue those externalities. Two examples of such cases are (a) the instruments to financially support renewable energy sources (RES) or for enabling their full integration into the electricity market, and (b) the design of capacity markets for system adequacy to be coupled with energy markets that are less and less reflective of the fixed costs of renewable electricity production.

It is, however, the last circumstance, the one relating to energy and climate transformation, which throws the most dramatic light onto our dilemma. In the times we are living, the energy context varies very rapidly and profoundly: several elements may be taken into account and they are both a driver and an enabler of the major transformation that is taking place. Typically, these elements are encapsulated in the well-known 3D slogan: decarbonisation, decentralisation and digitalisation. It could be said that the effects of 3D and the speed at which these effects are realised are not known, nor is it easy to obtain realistic prospects for the near future. As a result, one is left facing different and multi-faceted "uncertainties", as is the case also in energy regulation activity. Besides, those uncertainties undermine the solidity of the two first circumstances: the maximum information underpinning the decisions of the regulator and the predictability of the effects of the investments in an ever-changing context. Therefore, the aforementioned dilemma for the regulator reaches unprecedented new levels. Should I formulate the problem arising from that dilemma, I would try to formulate it in the following way: How to promote energy investments that also have a positive meaning in the future and how to do so in the present in a responsible way? Where to find the necessary 'guidance' to resolve this problem?

Whatever the answer to this problem, it is clear why it is crucial for the regulator to be as much enlightened as conscientious and pragmatic. And this is even more necessary than in the past.

2 Bit of Future-Proof Energy Regulation

If the reader in the previous paragraph was convinced that the energy regulator should develop a threefold character, that is, at the same time to be well informed, reasonably pragmatic, and necessarily enlightened, he should also be curious to know 'how' such character may be acquired. In other words, he may wish to know how to move from the "to be said" in formulating the easy equation of the dilemma to the "to do" in order to solve the difficult question. Or, at least, he may be interested in knowing how in reality regulators have attempted to promote investments in the energy context under awesome uncertainties.

But before bringing real examples, it may be useful to try and make a further synthesis of what we have just identified as the triple character. It seems to me that if regulators were "future-proof" in making regulatory choices, the desired synthesis of the triple character would be made. Future-proof posture is what is definitely needed.

Being future-proof for a regulator does not mean the ability to predict the dynamics that will materialize in the future (nobody can be required to have a certified crystal ball), but rather to include in the decisions he makes a sufficient, *quantum satis* amount of "insurance" that such decisions will have a "content of the future", so to withstand expected future developments. My experience as a regulator in energy has taught me that this type of insurance, albeit partial, is what is needed to allow investments to proliferate in the energy sector with reasonable and secure return and to avoid possible stranded costs.

Accordingly, these investments can be qualified as future-proof too.

Upon this narrative of what may be future-proof from a regulatory perspective, it is clear that the difficult part of any decision lies in the aforementioned *quantum satis* of content of the future (as in any recipe, it is not simple to blend an ingredient that has an unintended effect), and in the level of insurance for the future that the regulator is willing to adopt in order to promote investment in the sector (as with any insurance policy, there is a price to pay in the present to be shielded from a given risk in the future!).

There is, of course, no uniform and all-inclusive recipe for all types of investment or regulatory decisions to promote energy investment. This has to be done on a case-by-case basis, since the generalisation is so far impossible. The following paragraphs will deal with the *quantum satis* issue on a case-by-case basis, focusing on four examples coming from my past experience as energy regulator[1] in promoting future *investments*.

And in order to demonstrate the real difficulty of deciding on the future-proof and, above all, of behaving honestly, I have shown four examples of different nature. In particular:

1. The (historical) reform of the *wholesale natural gas market* (2012–2014) with the creation of price signals on the basis of economic merit with market mechanisms and the transfer of the resulting price benefits to all gas (and electricity) consumers, including households. This was an important decision, which has shown success.
2. The decision to renew the second generation of the *smart meter* fleet in the Italian electricity sector (2015–2017) in a large part of the territory served by distribution networks. This was really a future-proof investment which has proved successful in empowering consumers and the future self-consumers too.
3. The overall reform of the *Italian retail market* and, in particular, the excess of price protections in the electricity sector (2016–2018). The decision in some respects may be regarded as not being carried through for several reasons, even if this does not entirely depend on the regulator. Despite the positive development of innovative instruments (*Simile protection* and the *Placet* Offer), the partial completion is due—to a large extent—to the lack of sufficient future-proof criterion in the regulatory mechanisms because of persisting uncertainties concerning the future retail market design.
4. The (missed-) revision of the mechanisms meant to increase the *integration of RES* in the electricity sector (2017–2018), with the (non-) definition of innovative alternative tools to replace the incentive schemes for RES and manage different risks for market participants, including the 'renewable' purchase

[1]The experience referred to is relevant for the years 2011–2018 where the author was chairing a plural (collegiate) body (5 members) of ARERA, Independent Institution of Energy Regulation, based on a staff of over 250 officials involved in regulating the electricity and gas sectors.

obligations issued to consumers. This example is indicative of the lack of future-proof drive of the regulator, partly because of the end of the 7-year remit, partly because of some delay in having a European framework for new rules on the energy/climate transition (only available at the beginning of 2019).

3 Regulatory Future-Proof Actions/Mis-actions: Some Examples

3.1 The Future-Proof Reform of the Gas Market

During those times, I had repeatedly stated in public that the situation of the gas sector (far from being a market in the years 2010 to 011) could be likened to a "stone forest". The economics of that sector were those imposed by (historical) long-term take-or-pay contracts indexed to the oil price (ToP) and the regulator at that time was forced, albeit with concerns, to transfer administratively to consumers (especially domestic), the costs of those contracts that were completely out of the market when compared to European market prices, already characterised by liquid, competitive hubs. In short, Italy was reduced to "a gas province" where the commodity—despite having the same non-European origin—was paid much more than in Mitteleuropa. As a result, a situation of absolute lack of market signals persisted in the Country.

On the demand side (both industrial and mass segments), the situation was the same. The midstream importers of gas in the Country, who saw their costs (and margins) on the ToP contracts fully recognised by the regulator, operated price discrimination across market segments, benefiting from reduced prices to industrial customers (for the sake of preserving Italian firms' competitiveness at European level), while the other retailers competed on the retail market (for a policy that prevented them from being too vocals *in* pointing out the absence of effective competition). For electricity producers with gas-fired plants, the gas commodity was sold in such a way that profits could in any event be realised in the downstream electricity market, since combined-cycle gas installations were marginal in setting electricity prices in all hours of the electricity market. The increased costs in that "stone forest" were almost entirely borne by domestic consumers and small enterprises through their gas and electricity bills.

Having inherited this situation, we determined that the structure and economics of the sector should be completely rethought on the basis of two principles that are—of course—future-proof: the first one aimed at breaking the link between gas and oil prices, which was the core of the ToP contracts, given that gas in Italy (a sector which had peaked at 90 billion cubic meters per year in the first decade of 2000) was no longer a substitute for oil. We thought it was time for gas to be promoted only on the basis of the Country's supply and demand game and that the fate of the gas market in Italy should be the same as that of the other Member States of the European Union. If there should ever be a link to any commodity outside the country, this was to be sought with the most liquid hub prices in Europe (e.g. Zeerbrugge). Secondly, we determined that the system of different segmented price policies should be held back because an explicit wholesale gas price should be a single and transparent signal, known and visible to all classes of consumers. We started to create a market for balancing economic merit, based on gas offers for daily system balancing, with some fear (setting up an ad hoc "first" price monitoring system) that this newly-born gas market would be manipulated by the dominant players in order to discredit the emergence of market mechanisms in the eyes of consumers.

The reform gradually took place over a two-year period on the basis of consolidating the link of the Italian wholesale market (Virtual Exchange Point or PSV) with the North European liquid markets. Still today, the reform continues to show its benefits, on the one hand, by anchoring the average price of gas in Italy to that of Mitteleuropa, net of the costs of so-called logistics, namely the north-south transport of the gas commodity, and, on the other hand, by encouraging the specific local exploitation of gas in Italy because of the scarcity dynamics there.

The reader may object that I have not made clear the part of the promotion of future-proof investments following the gas reform of 2012–2014. One should then consider, for instance, two types of investment supported by the reform: (a) the reform initiated a large season of renegotiation of the incumbents' historic contracts in order to link them to the price conditions of the new Italian PSV market; (b) from the point of view of investments in the gas supply infrastructure in the Country, the reform provided the basis for attracting a number of new gas import infrastructure in Italy, both through pipelines (EU corridor, South corridor) and LNG (several regasification plants on the Italian coast). Moreover, it fostered investment aimed at enhancing the internal backbone of transport. It did so in a way that we now may think about the Mediterranean gas hub.

But, as we all know, if we want to create a gas hub, it is primarily necessary to favour an excess of building entry infrastructure in order for gas importers to compete with each other (and the regulator with its own pricing instruments may be decisive in the facilitation of the take-off). This approach carries the risk—everything to be calculated—that such over-investment may not be future-proof if the gas sector for the future is uncertain as to future demand and market developments because of the stranded costs to be borne by Italian consumers. The regulator was cautious in not going beyond, or even I would say, precisely: Future-proof.

3.2 *New Future-Proof Fleet of Smart Meter 2G*

Dynamic electricity pricing also exists in the mass market and Europe has rightly introduced it in its recent legislation. Dynamic pricing should allow for a 'punctual' differentiation of the value of energy for the retail market over time and territory, by diversifying the purchase and sale price of energy on the basis of parameters *hic et nunc*.

I recall that, in 2014, the regulator was confronted with several consultations (and many discussions) with stakeholders and consumer associations to reform the so-called "time-of-use cluster" registered by the consumers' first generation (1G) electronic meters we inherited from the previous regulation. 1G meters could only store three different bands per month where the hourly volumes of energy drawn by the consumer from the delivery point were aggregated within the cluster by providing a total monthly cluster value that had lost all information about the hourly consumption profile and all signals of hourly wholesale energy pricing. Very far from the concept of dynamic pricing of today! And not only that: with the three monthly clusters of 1G meters there was wide cross-subsidisation between consumers who paid energy according to an average category profile and did not receive any signal to shift their consumption (not even knowing their hourly profile). In addition to these huge inefficiencies, the fleet of 1G meters (over 30 million low-voltage connected customers) could not be efficiently used to cover remotely a range of commercial activities (e.g., change of power or closure of point of delivery) that today no longer makes sense to run administratively with traditional methods. In other words, the 1G fleet had been designed to collect monthly information from the meter fleet from the periphery to the distributor's processing centre, but in the opposite sense (centre to periphery meter) the 1G system bore only a few service messages from the distributor. To give an idea of the asymmetry in which we were, the entire fleet data could be 'downloaded' within one month, while to implement electronically via service messages a new time cluster system (which is only a weekly schedule divided into three sets of hours) would have required 24 months, even ignoring the likely unsuccessful attempts. Everything in 1G was implemented on proprietary technology and protocols of the distributor, leaving little room for improvement with the upcoming best available technologies.

In this case too, we opted for a strong future-proof decision in many respects, taking advantage of the approaching full depreciation of the 1G fleet, which would not have imposed any stranded costs for consumers. First of all, we designed a new set of functionalities that the second generation of 2G meters—or smart meters—would have to be equipped with. We were not in a position to choose the technology as we were well aware that it is not the regulator's job to indicate technological choices in lieu of network or market operators.

Four main building blocks of the 2G smart meters functional design were chosen because of their content of future: (a) record and make available to the final customer (or his representative) "the day after" and on an ad hoc consumption portal the time profile of the validated samples (i.e. checked and cleaned against any errors in the recording or transmission); (b) dimension reciprocally a substantial flow of data/commands from the centre to the individual meter on the field, comparable to the daily collection of measures of the entire 2G fleet; (c) opening up the 2G system protocols in such a way that they are defined and certified by standard technical bodies and that no proprietary exclusivity is established on them so as to ensure full interoperability of the 2G system (this latter characteristic is essential in the presence of a multi-distributor—approximately 150—and a multi-vendor context such as the Italian one, but also in sight of the approaching expiry of distribution concessions on Italian territory, which could see invitations to tender for the choice of distribution operators other than historical ones); (d) equipping the 2G fleet with an "internal" functionality for the final customer that can receive real-time measure (which has not been validated), from the meter with which its sampling point is equipped, of its withdrawal for the most diverse applications of home automation, data broadcasting and personalisation of commercial offers, including dynamic pricing.

On such a future-proof platform many applications are being developed both for customer data management by energy suppliers and for optimisation of their own consumption by final customers themselves and also of their own domestic production in the case of a self-consumer.

Let me to cite here a feature to which I feel particularly linked to, since I was pioneering in the energy sector: the so-called 'prepaid'. It is not the extension of the *local pre-payment (coins)* in force in the United Kingdom for vulnerable customers, but a mass market offer to have a preloaded account by the energy supplier in which the 2G meter is an essential link both to account for consumption in the account and to engage with the consumer as regards the credit status of its supply. The pre-payment will allow for a range of creative offers, including in the energy sector, and can be smartly used to reduce speculative claims (not linked to the vulnerability of the final customer) to the safe advantage of the reduction of counterparty risks in the market, and thus of prices, for the regular paying customer community.

However, the regulator's anxiety to launch such a fundamental future-proof project for the emancipation and empowerment of the consumer did not cease at the functional design stage but continued at the stage of approval of the first concrete roll-out plan for 2G smart meters compatible with the abovementioned functional design. One of the main distributors (active in 85% of the Italian territory) developed a solution—with an investment of over €5 billion for a 4–5 year installation campaign—based on a mature and highly established technology for the bidirectional transmission of data (power line carrier, or PLC) from the 2G fleet to the centre and vice versa. Precisely because it was very mature, it experienced from the beginning technological competitors who declared to be able to implement 2G features in a more reliable manner. Among its competitors, there were mainly telecommunications operators who, with the Narrow-band IoT technology, claimed to be chosen (although not yet commercially available, it was estimated to be so within 2 years from the date of the consultation), since they had even more future content than the mature PLC. The regulator, carefully designed to remain neutral in technological terms, but well aware to not rule out the future-proof characteristic of such an important investment, made an enlightened choice that (a) did not delay the approval of the roll-out plan of the new 2G fleet based on the smart meter 2.0 version to allow the release of benefits to consumers as soon as possible with PLC (b) remained flexible as regards technology by defining within the next 2 years a possible approval of version 2.1 (with additional communication technology) on the basis of field monitoring and testing on pilot projects of version 2.0 and its flexibility. In case it was necessary to upgrade to the 2.1 version the 2.0 version already installed in the field, a retrofit was already planned in order not to discriminate between segments of final customers. It is from these days, while I am writing, the decision of the present regulator that confirms the reliability of the 2.0 version (already about at half of a mass installation campaign) and the absence of need to update the 2.1 version. The future-proof worked! Before closing the paragraph, I would like to point out that the renewal of the 1G smart meter fleet to 2G has not involved any additional cost for the Italian consumer. Rather, using a simple car-related analogy, it has brought the smart metering system functionalities from those typical of a Trabant at the age of the Iron Curtain to those of a Ferrari type: all of this in line with the pricing invariance of the measure activity for the final consumer required by the regulator when approving the 2.0 plan.

What about the promotion of future-proof investments made possible by 2G smart meters? Abundant and large ones throughout the electricity value chain.

3.3 *Reform (Unfinished Future-Proof) of the Electricity Retail Market*

We know that the liberalisation of the energy market produced by Europe with its sequence of energy packages (from electricity of 1996 to the most recent *Clean Energy for Europe 2019*) has made gradual steps in making the energy market fully *eligible* and commensurate with the level of awareness of the final consumer when dealing with the markets. A good proxy to measure, at least on average, this level of awareness has always been identified in the 'size' of the final consumer. This process began with large industrial users, in order to switch to small business users. The last segment to be liberalised was the domestic one on 1st July 2007. Even in Italy, the transposition and the provisions of the regulator on the mass market put an end to the so-called regulated tariff for domestic customers at that time, opting for a market solution where the supply of energy was entrusted to a single central purchasing entity which supplied the entire portfolio of customers that—although eligible—had not opted for a supplier on the free market and a roll-over of the costs of this market portfolio on a quarterly basis with a publication of energy prices by the regulator. It did not, therefore, exercise any real price regulation (it was illegal and the European Commission—although it had opened a pilot in 2011 on that mechanism—concluded not to intervene on the ground of unlawfulness in the light of the liberalisation rules), but rather a kind of surveillance of prices independently negotiated by the purchasing centre on the wholesale market. Customers who considered that they had sufficient awareness to face, individually or in purchasing group, the dynamics of the markets could voluntarily abandon this mechanism by virtue of their rights as an eligible consumer. Several discussions and consultations facilitated by the regulator sought to assess the future-proof of the protection model (I repeat, only at the level of surveillance of the market prices specified therein), by collecting opinions that are almost always in line with (legitimate) interests, but never considerations relating to the need to overcome this model.

The least future-proof part was raising (and still raises, when I am writing) significant problems of blocking—or at least narcotising—market liberalisation was (and is) the one linked to the actors who, heirs of the old concessions for distribution and sale of pre-liberalisation energy, were (and are) exclusive holders of the provision of the sale service to customers who do not leave the surveilled price protection (also called "greater protection"). This arrangement was established by an Italian law of 2007 and has never been amended ex lege. A number of radical solutions have been proposed (e.g. Customer Batch auctions) to collectively move to and forcibly migrate into the open market so-called 'inert' end customers (i.e. those who neither had done, nor intended to do, a voluntary switch), to tell the truth, all of which did not respect the free determination of the final customer who, at a closer look, underpins the European principles of liberalisation of the retail market.

To try to give confidence in the market and awareness of their rights and market mechanisms, the Regulator introduced (2015) the so-called "Simile Protection" and (2017) the *Placet* Offer (in Italian: Prezzo Libero A Condizioni Equiparate di Tutela), which were subsequently supported by a special law known as "2017 Competition Act"). The latter are two forms of enhanced protection of the inert consumer without any limitations as regards sales prices: The Simile Protection is an offer 'at a discount' in respect of the monitored price conditions while the *Placet* Offer (Free Price At Equal Conditions of Protection) is an instrument which must be offered by every seller, who addresses the mass market with a fixed and standard contract (approved de facto by the regulator) but with a price freely set by the seller. Future-proof tools in the retail market also play an important role in empowering inert consumers to let them switch from the greater protection service and be served from market suppliers.

However, given the uncertainties of the future market structure as regards inert consumers, which is not specified by the Government to which the 2017 Competition Act referred the definition task (competence not yet exercised as I write), the regulator considered too risky—and thus not sufficiently future-proof—forcing the issue and defining the coming structure on his own. Even if the preferred model for the inert, we can now say, was that in which the commodity purchasing centre remained in place, in order to exercise market price control over it, although supervised by the marketing service to the final customer (contact and relationship with the consumer, billing, invoicing, customer care) offered to tender in batches to market operators or even new entrants in the sector.

This last failure to evolve is the one that I call "unfinished" reform: not because of the lack of information and pragmatism of the regulator, but rather due to the absence of the necessary enlightenment on future market structures as desired by the Government.

3.4 *(Missed) Review of the Future-Proof Integration of Renewable Sources*

It would be unfortunate to say that the Italian regulator in 2011–2018 did little for the integration of RES, especially in the electricity market. It is true, however, that in comparison with the ambition of the objectives that had to be defined, integration was not pushed to the appropriate level because of the many hesitations stemming from non-positive past experiences and those induced by the evolving regulatory framework. Why do I claim these two reasons?

Because the 2008–2017 season to incentivise renewable electricity was—rightly—perceived in Italy as a too generous experience in terms of incentive levels set by the Government with very noticeable effects on all consumers' bills. Apart from the fact that such incentives were largely defined by the policy maker rather than the regulator, it is undeniable that the public policies introduced have led to rather unsustainable burdens on the so-called general system charges on electricity bills (peaking at around €14 billion per year), which led to:

(a) a price gap between the wholesale electricity market—characterised by a significant price reduction during the hours of massive solar and wind turbine production at no variable cost—and the retail market that is charged to the administered incentive to recover incentives. Thus, the competitive pressure on the commodity in the latter market was also reduced (for the average representative domestic user the commodity-tied part of the commodity market fluctuated in those years around 40% of the total bill spending, the remainder being around 15% for taxes and, in equal shares, 22% for general system charges and network charges). This led to the disappearance of the market pricing of part of the production costs in the wholesale market and their conversion into compensation administered as an incentive;
(b) given the situation of Italian public finances and the relative difficulty in charging general taxation with part of the burden, charging the entire burden of the incentives to consumers of the electricity sector has discouraged further penetration of the electricity carrier by comparison with other (fossil) sources that have a higher impact from the point of view of climate-changing gas sources;
(c) the need to alleviate from this heavy burden—at least in part—energy-intensive industrial sectors in order to maintain their competitiveness in comparison with European competitors. This has been done in compliance with European rules for the protection of public funds which may constitute State aid, but certainly is the direct consequence of renewable incentive schemes;
(d) the incentive to escape from the payment of the general system charges which, being configured as increases in transport charges for energy taken from public networks, could be 'avoided' by favoring all forms of viable exemption (e.g. net metering) and self-produced energy and self-consumption schemes, which, as is well known, avoid subjecting energy to general system charges.

However, this means, irrespective of whether self-consumption is for the production of renewable energy, the creation of *enclaves* free from such charges and the resulting increase in the unit rate of collection of such charges for the rest of the user community not included in the aforementioned *enclaves*;

(e) the occurrence of high system balancing costs compared to the previous periods, as the aleatory productive nature of solar and wind RES entails, for the maintenance of the safety of the electricity network, an increase in the so-called reserves and of the balancing energy purchased from the system (such as increasing or decreasing services in relation to the planned profiles), failing to provide efficient and adequate storage facilities within the electricity system.

The second reason is even greater uncertainty because of the hesitations stemmed from the debate about the more or less ambitious objectives that were being defined for the development of RES and that are only now being answered in the European Governance process through the Integrated National Energy and Climate Plans (Regulation No 1999 of 2018).

For these reasons, the discussion within the regulatory body in 2016–2017 was affected by many uncertainties that were, in turn, based on the negative lessons learned from past experience (see above) as well as on the costs of RES that did not show a significant reduction (as would later happen from 2018 onwards). This situation was further compounded by vagueness about credible objectives for the promotion of renewables and the ways to achieve them (in particular, we were wondering if the national policy maker aimed at utility-scale renewable plants or spread on the territory or at both).

As a result, the discussion in the regulatory Authority could not be kept informed, or pragmatic, given the negative lessons learned, and much less enlightened because the policy maker was slow in setting objectives and trajectories, enabling the regulator to build the appropriate instruments for their achievement.

It was, therefore, implicitly decided that it was not a case of "anticipating" an energy policy that has not yet been set out by regulatory instruments. In fact, for our purposes, it was decided to overlook the issue as the future-proof test had not given any green light yet.

4 Promotion of Future-Proof Investments: A Never-Ending Regulatory Regulatory Saga

In order to promote future-proof investments, every national and European regulator (if there ever will be one) must 'invest' in its ability to be future-proof and make decisions '*quantum satis*' future-containing in order to ensure that their regulatory choices are resistant and resilient enough to withstand likely future developments in the sector. Being future-proof for a regulator does not mean to make choices regarding specific technological solutions because the typical profession of a regulator is not that of the technologist. On the contrary, his task is to design and implement market instruments and infrastructure frameworks for the purpose of achieving general interest objectives that would not be achieved otherwise. Rather, the regulator must work in a future-proof way on functionalities, behaviors, and mechanisms. If he/she did not do so, even if within the limits—even strict—of his/her capacity to be fully informed, usefully pragmatic, and necessarily enlightened, he would affect the adequacy of the regulation for future-proof investments or, on the contrary, there would be a panoply of non-future-proof investments.

So the regulator's dilemma continues, or better it becomes a never-ending regulatory saga.

Acknowledgements The author acknowledges the valuable contributions to the present chapter provided by Marion Malafosse and Federico Gallo.

Disclaimer The information and views set out in this chapter are those of the Author and do not necessarily reflect the official opinion of the Author's current affiliation with the European Commission.

Part II
Energy Transition, Business Transformation, and the Role of International Players

From National Champion to Global Player

Francesco Starace

Abstract This paper aims to represent how Enel evolved from a national champion to a global leader in the Energy Transition, highlighting the paramount role of renewable energy, innovation and sustainability. Enel made sustainability the engine of its innovation model, not only to face today's challenges but also to create sustained long-term value for the company and for all its stakeholders (e.g. investors, customers, communities, employees, etc.). In November 2017, Enel launched Enel X, a new global business line, aimed at opening up the energy sector to new business models and new technologies that can capture the value opportunities emerging from the evolution of the market. Moreover, this paper showcases how Enel has chosen to take the lead in the transition towards a more sustainable model with Futur-e, designed as a circular economy pathway to bring 23 decommissioned thermoelectric power plants and a disused mining area back to life through direct and active collaboration with local communities. These new eco-sustainable areas will be dedicated to science, art, culture, tourism or even new industrial activities.

Keywords Enel group · Energy transition · Renewables · Sustainability · Innovation

1 Energy. From National Champion to Global Player

1.1 The Energy Sector at the Root of the European Integration Process

The Italian electricity industry was born at the beginning of the 20th century and has lived through many major historical events, including two world conflicts, the creation of the European Coal and Steel Community in 1951, the establishment of the European Economic Community in 1957 and finally the European Union (EU). Since 1962, the year when Enel was created, the Italian electricity industry has evolved along with the history of the Enel Group.

A. Gilardoni (ed.), *The Italian Utilities Industry*,
https://doi.org/10.1007/978-3-030-37677-2_4

The ambitious path towards European integration, which was intended to ensure peace and prosperity on the continent in the aftermath of the Second World War, rested on two pillars that formed the European energy system: production, largely based on European coal resources, and industrial manufacturing, in which the steel industry played a strategic part, as it still does today.

The founding fathers of Europe understood that integrating and commonly managing their heavy industries—coal and steel—was a way of contributing to peace among nations, preventing an uncontrolled rearmament, as well as being the first step towards the economic, political and social relaunch of a continent that had always been poorer in primary energy resources compared to the major competing economies.

1.2 The Creation of the European Electricity System: Enel Before the Bersani Decree

The Italian electricity industry has a long history and its main players were initially private entrepreneurs, organised into companies specialised in generating electricity or businesses which, despite operating in other sectors, considered it economically convenient to self-generate the electricity they needed. Most of the energy required at the beginning of the last century was provided by hydroelectric power, thanks to Italy's particular orographic structure.

Thanks to the cooperation between captains of industry and financial institutions in the 1920s the production of electricity, in particular hydroelectric power, grew considerably. The electricity companies and their investors were hit hard, however, by the Great Depression. The economic situation sorely tested major producers like Società Idroelettrica Piemontese (SIP) and it was only thanks to State intervention that the Italian electricity industry overcame the difficulties and managed to withstand the economic crisis.

In the 1950s, there were hundreds of electricity companies in Italy, more than half of which were controlled by only six large groups. In the north, from west to east were SIP, Edison and Società Adriatica d'Elettricità (SADE). Southern Italy had Società Meridionale di Elettricità (SME), while in Sicily and Sardinia electricity generation and distribution were guaranteed by Società Generale Elettrica della Sicilia (SGES) and Società Elettrica Sarda (SES). Alongside private companies, there were also municipal companies in big cities like Turin and Milan and self-generators.

In 1962, in line with the economic vision of the time, the Italian government launched the nationalisation of 1270 local electricity companies, thus promoting the birth of Enel, then known as Ente Nazionale per l'Energia Elettrica.

Enel was a public entity that operated, however, similarly to a private company, both in its relations with its employees and in its activities, and was required to be profitable, seeking to achieve cost efficiency, while operating in a non-competitive regime.

The objectives of the newly established Enel were very challenging: the first years were spent unifying voltage levels across the national electricity grid, while at the same time streamlining operations to deal with an expanding economy and population growth during the years of the so-called "economic miracle". Enel also immediately began to invest heavily in improving the quality of the service and adapting its generation capacity to support the Country's transition from a rural to an industrial and finally a service economy.

Between the 1960s and the 1980s, Enel built hydroelectric power plants that still generate electricity today: San Fiorano, with a total capacity of 568 MW, Edolo (1000 MW), Luigi Einaudi (1065 MW), and finally Domenico Cimarosa (1000 MW). At the same time, while enhancement work continued on geothermal production plants like Larderello in Tuscany (already operating at the beginning of the 1900s but further expanded by Enel), Enel undertook the construction—or in some cases boosted the capacity—of the large thermoelectric power plants in Rossano Calabro in Calabria, Fusina and Porto Tolle in Veneto and Brindisi in Puglia (Figs. 1 and 2).

At this time, direct intervention by the State throughout the value chain became a characteristic of the electricity industry, not only in Italy. This often involved the creation of vertically integrated State-owned monopolies with the aim of guaranteeing security of electricity supply. During the 1980s, however, a reflection process on a possible gradual liberalisation process began, reaching its maturity in the early 1990s in Europe. The dominant idea was that, even in the electricity sector, a widespread and efficient competitive market model had to be introduced, returning in some respects to the private entrepreneurship model from which the industry had first emerged.

The European liberalisation process went through three main "Energy Packages", with the first one issued in 1996 (the European Directive 96/92/EC concerning common rules for the European internal electricity market), and the others following, respectively, in 2003 and 2009. These legislative frameworks provided for:

a. the liberalisation of the electricity generation activities, previously characterised in many countries—including Italy—by an exclusive concession regime, and the creation of wholesale markets, first at national level and then increasingly integrated across Member States;
b. the gradual process of opening up the retail electricity markets, giving to customers the possibility to choose their own supplier;
c. an increasing separation of the activities of vertically integrated companies (so called unbundling) between, on the one hand, supply and production activities (where competition is possible) and, on the other hand, network operations (which are a natural monopoly), with the latter subject to regulatory mechanisms intended to guarantee non-discriminatory access to essential infrastructures for all operators;

Fig. 1 Fumaroles in Larderello, Tuscany. *Source* Enel Green Power photographic archive

Fig. 2 Sasso 2 geothermal power plant: cooling tower, Larderello, Tuscany, 2014. Photo courtesy of Fabio Sartori

d. the introduction of independent National Regulatory Authorities (NRA), to ensure non-discrimination, effective competition and efficient functioning of the market.

During the years in which the liberalisation process took place, the majority of the former public monopolies has been privatised across Europe.

1.3 The Italian Electricity Market: Enel After the Bersani Decree

Italy transposed the European Directives into national law, radically transforming the competitive structure of the electricity industry in relatively short time through several interventions designed to have profound and lasting effects on the market structure and its operators:

a. implementing the liberalisation process in Italy led to the separation of Enel's activities, while ensuring the Group's ability to play an important role in the international competitive context. In particular, Enel was required to sell: (i) 15 GW of its production capacity (for comparison, close to the total installed capacity of Greece) through so called Generation Companies (GenCos); (ii) its distribution networks in large cities where the national operator coexisted with local utilities (Rome, Milan, Turin, etc.); (iii) at a later stage, its own transmission network, which was transferred to a separate company, Terna. At the same time, the regulatory framework was designed to support the creation of a competitive arena, making the Italian market one of the most competitive in Europe;
b. allowing international operators to enter the market by acquiring the generation capacity sold by Enel, thus creating an embryonic supranational electricity market.
   Opening the market to foreign companies, together with increasing generation investment made by other Italian companies, brought to a drop of Enel's market share in the Italian electricity generation market from over 70% in 1998 to around 19% in 2018[1];
c. establishing a Regulatory Authority (Law No. 481 of 14 November 1995) with strong and independent powers, only matched at European level by the already mature British experience.

The way in which the electricity industry has evolved has played an extremely important role in the history of major electricity companies and Enel is no exception in this respect.

[1]Annual report on services and activities, 04 July 2019.

For Enel, the start of the liberalisation process coincided with a radical rationalisation of the organisation and the start of a process which led, in 1999, to the company's capital being opened up to private investors with the listing of 30% of its capital simultaneously in Milan Stock Exchange and on Wall Street.

Enel responded to the reduction of its market share in the Italian electricity market by implementing two strategic actions: diversification and subsequently internationalisation.

Diversification saw the Group expand into contiguous sectors according to the so-called multi-utility model, an approach that gained strong support in the 1990s. It entered the telecommunication sector by developing Wind, which was granted its first licence in 1998, and acquired Infostrada in 2001, consolidating the Group's presence in this sector (reaching over 20 million customers). During the same years, Enel launched its first water distribution activities. This phase ended a few years later, at the end of 2005, with the sale of Wind and the water distribution business and the company returning to focus on the electricity sector in Italy and abroad.

Utilizing a large part of the capital freed up by the disposal of non-core assets, the Group began to expand internationally with increasing determination. Its overseas operations began in 2000 with the acquisition of CHI Energy, a renewable energy producer operating in North America, and in 2001 with the acquisition of Viesgo, an electricity generation and distribution company operating in the Spanish market. The entry into Eastern Europe (Slovakia, Romania, Russia), which began in 2004, was instead the natural consequence of the liberalisation and concurrent privatisation processes taking place in these countries.

The real turning point for Enel was the acquisition of Endesa, which began in 2007 and was completed two years later. This operation allowed Enel not only to incorporate the leading Spanish electricity company, but also to access the South American markets, a region characterised by high rates of economic and demographic growth, with an important energy demand to be satisfied. With these operations, within a relatively short period of time, Enel became a global player, ranking second among European utilities in terms of installed capacity.[2]

By the end of the first decade of 2000, this internationalisation process had transformed Enel from a state monopoly into a major international group, a world leader in the sector, with a diversified, balanced generation mix and a presence in many foreign markets.

1.4 *The Centrality of Environmental Issues and the Extraordinary Development of Renewables*

Viewed retrospectively, the liberalisation process can be considered a success in many respects: competition has led to increasing investments that have advanced

[2]Enel, Annual Report 2008.

the performance of thermoelectric power plants and spreading technologies aimed at improving the electricity system as a whole.

These benefits have been particularly significant in Italy, where the installation of over 30 million smart meters (further information available in dedicated chapter on E-Distribuzione), almost twenty years ahead of the deadline set by the European Union (2020), has contributed to improving the quality of the service. Likewise, the modernisation of generation facilities through the application of the best available technologies has brought significant benefits, both environmental and economic.

Initially, however, the liberalisation process was unable to identify appropriate solutions to two other major challenges: sustainability and energy security. From the 1990s onwards, these issues started to be addressed and became subject to major multilateral agreements, including above all the Kyoto Protocol, aimed at reducing Greenhouse Gas (GHG) emission in the atmosphere, promoting the decarbonisation of the economic system and the development of new sustainable technologies. The Protocol established binding quantified emission limitation—or reduction commitment—only for developed economies and economies in transition. The Protocol also established non-binding limits on GHG emissions for emerging economies based on the principle of common but differentiated responsibilities and respective capabilities. Under the Protocol, developed economies have a greater mitigation role than emerging ones, but this principle showed its limits when the overall emissions by emerging economies rose dramatically. This led to a growing concern among developed economies that believed the agreement was ineffective, thus did not ratify it.

Given this experience, ahead of the Copenhagen Conference in 2009, Europe applied the "leading by example" philosophy, unilaterally pursuing the objectives set by the so-called "20-20-20" package, which established the following targets for 2020:

1. 20% cut in GHG emissions (from 1990 levels);
2. 20% renewables on EU final energy consumption;
3. 20% improvement in Energy Efficiency (EE).

Through these targets, which were translated into specific national objectives for all EU Member States, the EU intended to boost its decarbonisation process through renewables and EE, while reducing its dependence on imported fossil fuels.

The actions taken nationally to implement the 20-20-20 targets have seen several Member States—including Germany, Italy and Spain—launch very aggressive investments in renewable energy sources. These policies—initially administratively-set and then market-based—have contributed to the development in Europe of a new energy paradigm, cost reductions of renewable technologies[3]

[3]BNEF NEO 2019: solar photovoltaic is increasingly competitive as the cost of a utility-scale generation has fallen by 85% between 2010 and 2019 and a further fall of 63% is expected by 2050 (LCOE has fallen to $57 per MWh in 2019 from $375 per MWh in 2010 and is expected to be $21 per MWh by 2050).

and the development of a specific technological and business know-how, factors which have allowed European companies to operate successfully in Europe and abroad.

Moreover, in 2015, the Paris Agreement was adopted aiming to strengthen the global response to the threat of climate change by keeping a global temperature rise this century well below 2 °C above pre-industrial levels and to pursue efforts to limit the temperature increase even further to 1.5 °C. To achieve these goals, the Paris Agreement requires all signatories to undertake efforts towards reaching global peaking of GHG as soon as possible and towards achieving a balance between anthropogenic emissions by sources and removals by sinks in the second half of the 21st century. In order to overcome the limits of the Kyoto Protocol, the Paris Agreement required all Parties to define and communicate binding post-2020 climate actions, known as NDCs (Nationally Determined Contributions), reflecting the highest possible ambition under the principle of "common but differentiated responsibilities and respective capabilities". NDCs embody efforts by each country[4] to reduce national emissions and adapt to the impacts of climate change. In addition, the Paris Agreement invites all non-Party stakeholders (e.g. private sector, financial institutions, cities and other sub-national authorities) to scale up their efforts in order to lay the foundation towards the low carbon economies of the future.

1.5 Enel Today: A Global Leader of the Energy Transition

Undertaking a further radical transformation to fully embrace the development of renewables, Enel created in 2008 Enel Green Power (EGP). Over the course of a few years, EGP has become a leading global player in the development and management of generation assets from renewable sources. Today EGP is one of the main private operators in the world in terms of managed capacity, with a global presence on five continents. EGP manages almost 46 GW from wind, solar, geothermal and hydropower.[5] This process has brought countless benefits, not only to the company and its shareholders, but also to the Italian supply chain, which has been able to scale up and take advantage from Enel's business development worldwide.

Along with a further increase in renewable deployment, a fundamentally important area of development for Enel lies in the massive application of digital technologies to assets (e.g. generation, distribution grid, etc.), customers (e.g. improving relations and quality of service) and employees (e.g. improving digital skills). For example, a digitised grid (through the introduction of smart grids, further information available in dedicated chapter on E-Distribuzione), can integrate

[4]One single NDC submitted for the European Union on behalf of all 28 Member States.

[5]Data as of 27 November, 2019.

and manage more efficiently an increasing number of distributed renewable plants. Furthermore, the ability to manage the enormous amount of information coming from a digitised grid is creating a unique opportunity in terms of operating costs reduction, customer relations improvement and a broader range of services.

Given the growing penetration of renewable sources, it will be increasingly important to develop solutions that increase the flexibility of the electricity system, such as demand response, storage and Vehicle-to-Grid (V2G). In order to manage these and other innovative solutions, Enel created in 2017 Enel X: a global business line dedicated to the supply of new services to customers, including charging infrastructure for electric vehicles, distributed renewable generation, smart public lighting systems, innovative financial services, in addition to the previously mentioned flexibility tools for the electricity system (Fig. 3).

In just 15 years, Enel has been able to transform itself from a national monopoly into a global player and leader in the Energy Transition. A top European utility by market capitalisation and among the largest in the world, Enel is today synonymous with innovation, sustainable development and quality of service.

Today, the Enel Group operates in 33 countries across 5 continents, generating energy through a managed capacity of around 89 GW, distributing electricity over a network of approximately 2.2 million km. With 73 million end users in the world, Enel has the largest customer base in comparison to its European competitors and is among Europe's leading electricity companies in terms of installed capacity.

Fig. 3 Enel X's JuicePole electric vehicle charging station, able to recharge two vehicle at the same time

2 Integrating Sustainability with Innovation

For Enel, sustainability is an innovation driver. In the dynamic scenario of the Energy Transition, the limitations of a "closed" model—in which companies and individuals generate and develop ideas in an isolated system—have soon become evident. This is the reason why Enel has made "open" innovation one of the fundamental levers of its strategy, thus combining agility and high standards of efficiency. Taking a further strategic step forward, Enel has decided to combine open innovation with sustainability into "Innovability", in which sustainability drives innovation, based on the firm belief that in order to be sustainable it is necessary to constantly innovate and at the same time every innovation should be sustainable. Creating shared value in the long-term means abandoning a self-referential dimension and developing an increasingly inclusive relationship with all stakeholders, adopting an open, innovative, proactive approach to generate measurable benefits for the company and for the community.

The concept of "Innovability" integrates perfectly with the targets Enel has set itself for contributing to the achievement of the Sustainable Development Goals (SDG) set by the United Nations. In particular, Enel has centred its strategy around the achievement of SDGs across all of its activities, with SDG 13 on Climate Action as the cornerstone of the strategy. As a result, the Group engages in Decarbonisation of both production and consumption, while pursuing Electrification of end uses to tackle climate change as well as providing access to affordable and clean energy, in line with SDG 7. Key enablers of decarbonisation and electrification are Infrastructures and Networks, in line with SDG 9 on Industry, Innovation and Infrastructure as well as Ecosystem and Platforms, in line with SDG 11 on Sustainable Cities and Communities.

Sustainability for Enel also means circular economy, which already plays a strategic role in achieving a balance between environment and economic development, while allowing new business models to emerge. The adoption of a circular approach requires a radical rethink of companies' business models, from supply chain, through production processes to customer relationship. At Enel, circularity is already a feature of various applications, for example in tender procedures, where reward schemes are provided for suppliers who commit to reduce the consumption of natural resources and the production of waste over the entire life cycle of a product or service.

Collaborations within and beyond the energy sector play a key role, allowing Enel to scout for innovative solutions. It is also for this reason that Enel has adopted a true "Open Innovability" ecosystem, enhancing the entire value chain, in particular through a network of Innovation Hub and Lab, both Italian and international, in geographical areas with the highest innovation rate in the world (San Francisco and Boston, Tel Aviv and Haifa, Moscow, Madrid, Santiago, Rio de Janeiro, São Paulo, Pisa, Catania and Milan) (Fig. 4).

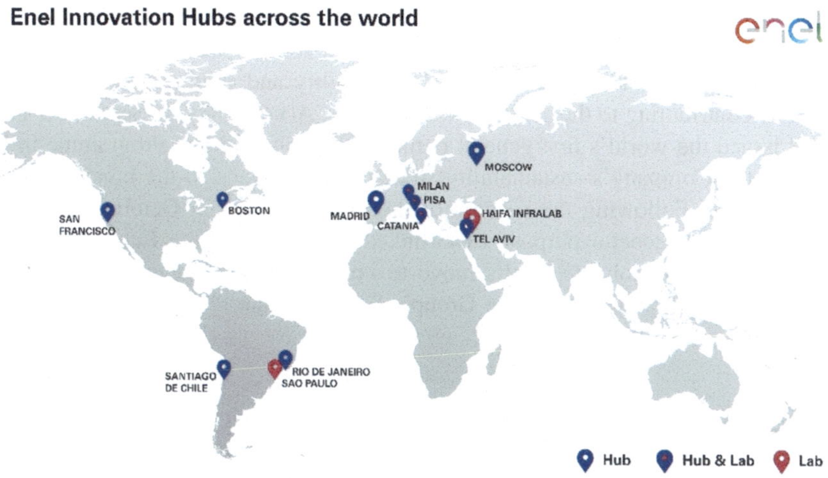

Fig. 4 Enel's innovation network across the world

A crowdsourcing platform has also been launched, thanks to which business challenges—linked to SDGs—are shared with an ecosystem of "solvers" (around 400,000) to find possible solutions. Moreover, Enel is exploring the use of hydrogen to enable the deep decarbonization of the economy.

In Enel's vision, if hydrogen is produced in a sustainable way (i.e. with a carbon-neutral footprint and from renewable sources), it can complement direct electrification to deliver a fully decarbonised economy by mid-century. Once the full efficient potential of direct electrification has been exploited, sustainable hydrogen can be used to tackle emissions in "harder to abate" sectors, like industrial sector requiring hydrogen as feedstock (e.g. ammonia and steel) or high-grade heat, or some heavy transport segments (e.g. long-range aviation and shipping). Enel believes that the only form of hydrogen that is truly sustainable is hydrogen produced from water electrolysis powered by renewable electricity.

To reinforce Enel's Commitment to sustainability and innovation, since 2004 the Group has become part of the United Nations Global Compact, that reconfirmed the Group as a LEAD company (a title it holds since 2011). Enel's global sustainability leadership is also acknowledged by the fact that it has received for the first time an "AAA" rating by MSCI ESG Research Ltd., a leading research and data provider measuring companies' performance on the grounds of Environmental, Social and Governance (ESG) factors. Moreover, Enel is also present in several renowned sustainability indices, such as the Dow Jones Sustainability Indices, the FTSE4Good Index, the Euronext VIGEO-EIRIS indices, the STOXX Global ESG Leaders indices, the OEKOM "Prime" rating, the ECPI indices, and the Thomson Reuters/S-Network ESG Best Practices Indices. Recently, Enel was named Europe's innovation champion at Corporate Startup Stars Award 2019, a recognition for the concrete actions taken on open innovation.

Finally, Enel believes that there is a clear link between sustainability and value creation as, by investing in environmentally and socially sustainable projects, the Group has been able to secure higher profitability and minimise its risks, while directly contributing to the achievement of SDGs. Accordingly, in September 2019, Enel issued the world's first general purpose SDG-linked bond Programme in US, linking the Company's sustainability strategy performance to the bond envisaged performance. Following the success of the first emission, in October 2019, Enel issued the first general purpose SDG-linked bond also on the European market. These actions consolidate Enel's move to implement an increasingly sustainable finance strategy, aligned with the Group's Strategic Plan.

3 Circular Economy from Theory to Practice: Futur-e

Managing the Energy Transition also involves defining a strategy for production sites, which in many cases have made the history of the Italian electricity industry, but that have fulfilled their generation mission and must now find a new purpose. In this respect, in 2015, Enel launched the Futur-e project, an internationally unique programme originally designed as a circular economy pathway to bring 23 Italian thermoelectric power plants, as well as the former Santa Barbara mining area, into eco-sustainable areas dedicated to science, art, culture, tourism or even new industrial activities, with the direct and active involvement of local communities.

Over time, the approach pursued in the Futur-e project has proved successful and will therefore be extended to geographies where the Group will be managing thermoelectric plants through the Energy Transition, such as Spain and Chile. In the current context, redevelopment opportunities will be expanded, including through the replacement of some of the 23 thermal plants with new renewable, renewable hybrid or gas plants capable of ensuring stability of the system during the transition towards a fully decarbonised economy.

4 Conclusions

The decarbonisation of the electricity sector has gone further than in other industries and for this reason the sector is increasingly being seen as a forerunner: technological changes and the consequent working skills evolution are manifesting themselves well in advance compared to other industrial sectors. To ensure that these developments are successful and embraced by societies, the very important concept of a *Just transition* must be applied, which involves:

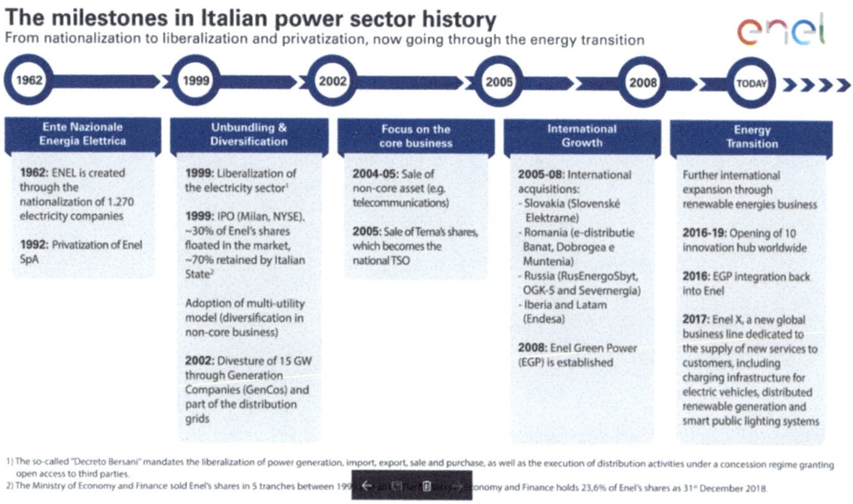

Fig. 5 Timeline showing some of the most important events in Enel's evolution

- supporting a fair Energy Transition for workers and communities through a greater integration of the social dimension with energy and climate policy;
- identifying the most impacted contexts and working with decision-makers to allow, for example, targeted training for workers whose jobs are destined to change in the near future and for those entering the labour market;
- fighting energy poverty by ensuring a universal access to clean, modern and affordable energy services;
- lastly, not forgetting that even companies must equip themselves in this constantly changing scenario, endowing themselves with agile and flexible forms of governance.

The change taking place is therefore an opportunity, but in order to be truly such, it is essential that the Energy Transition is managed gradually, ensuring that nobody gets left behind. Electricity is and will increasingly be the essential form of energy for sustainable development, a crucial element for ensuring the competitiveness of the Italian and European industrial value chain (Fig. 5).

From Letta and Bersani Decrees to the Future Challenges. The Role of Edison

Nicola Monti

Abstract Edison is Europe's oldest energy company and one of the largest utilities in Italy. Since the '90s the Company has played a leading role in the energy sector, by investing in the most advanced and clean technologies and providing a key contribution to the liberalisation of the Italian electricity and gas markets. Today, Edison's vision of the future relies on sustainable power generation, energy efficiency and high-value relationship with customers and territories. Indeed, the Company has constantly upgraded and innovated its generation fleet, expanded its renewables portfolio, and consolidated its expertise in the gas and electricity supply chain in order to support the energy transition until sustainable end uses like mobility. Moreover, following the liberalisation process, Edison has developed a wide range of energy products and services, increasing its presence in the downstream markets–creating shared value for the local communities–while continuously improving the efficiency and sustainability of the upstream phase.

Keywords Energy transition · Energy efficiency · Wind power · LNG · Wholesale energy market · Creating shared value · Edison group · Sustainability

Edison is a leading Italian energy company with more than 130 years of history, operating in the electricity and gas markets, with a strong geographic focus on Italy and Europe and Eastern Mediterranean region. At the end of 2018, Edison employed over 5000 people and served more than 1.5 million customers, with revenues of more than €9.1 billion and Ebitda of €793 million (see Fig. 1 for Edison Financial Performance).

In 2018, the company had a net installed capacity of 6.1 GW, of which 1.7 GW from renewable energy sources (65% hydroelectric energy, 34% wind power, 1% from other renewable sources).[1] On gas, Edison held four long-term supply contracts for a volume of 14.4 Gcm. Overall, Edison had a balanced portfolio of assets, which was able to meet 7% of power demand and 21.8% of Italy's total gas imports.[2]

[1]Edison (2019), 2018 Financial Report. Report on operations.
[2]ARERA (2019), 2018 Relazione annuale.

A. Gilardoni (ed.), *The Italian Utilities Industry*,
https://doi.org/10.1007/978-3-030-37677-2_5

Income statement highlights (in millions of euros)	2018	% of revenues	2 017 (**)	% of revenues	% change
Sales revenues	9,159		8,783		4.3%
EBITDA	793	8.7%	803	9.1%	- 1.2%
EBIT	199	2.2%	42	0.5%	n.m.
Profit (Loss) attributable to Parent Company shareholders	54		-176		n.m.

Balance sheet highlights (in millions of euros)	12.31.2018	12.31.2017	% change
Capital expenditures	418	377	10.9%
Investments in exploration	29	80	- 63.8%
Net invested capital (A + B)	6,557	6,319	3.8%
Net financial debt (A)	416	116	n.m.
Total shareholders' equity (B)	6,141	6,203	- 1.0%
Shareholders' equity attributable to Parent Company shareholders	5,886	5,915	- 0.5%

Key Indicators	12.31.2018	12.31.2017	% change
Debt/Equity ratio (A/B)	0.07	0.02	
Gearing (A/A+B)	6.3%	1.8%	
Number of employees (1)	5,372	5,144	4.4%

(1) Year-end data for companies consolidated line by line.
(*) See the Notes to the consolidated financial statements.
(**) 2017 Sales Revenues were restated following the application of IFRS 15 "Revenue from Contracts with Customers", without any impact on the EBITDA, as described in the section 1.1 "Newly applied standards" of the Notes to the consolidated financial statements. The effects resulting from the initial application of IFRS 9 were recognised in shareholders' equity with no restatement of comparative data.

Operating data	2018	2017	% change
Net production of electric power (TWh)	18.8	19.7	(4.8%)
Sales of electric power to end users (TWh)	13.7	10.9	25.3%
Gas imports (Bn m3)	14.6	15.1	(3.5%)
Total net gas sales in Italy (Bn m3)	20.7	21.3	(2.7%)
Power and gas contracts (thousands)	1,592.4	1,059.3	50.3%
Energy services contracts	419	391	7.2%
Hydrocarbon reserves (Mboe)	209.1	224.0	(6.7%)
Net hydrocarbon production in Italy and abroad (Mboe)	18.1	17.0	6.5%

Fig. 1 Edison consolidated financial statement 2018

1 Power Generation: The Contribution to the Efficiency of the Italian Fleet. The GenCo's Experience

Since 2006, Edison has been including energy efficiency among the cornerstones of its core vision as a key enabler for a sustainable future. However, Edison's focus on energy efficiency as a strategic priority had started much earlier. Indeed, in the '90s, Edison was the first operator to introduce in Italy combined-cycle gas turbines (or CCGTs), which stood out as the most efficient technology at the time, with a limited environmental impact in terms of polluting emissions.

Later on, Edison took part in the auctions organized for the sale of Enel's GenCos (i.e. generation companies). In fact, based on the provisions of the Bersani Decree (1999), the former monopolist had been legally obliged to keep its production below the 50% limit with respect to the Italian generation market, implying the divestment of part of its fleet. At first, about 15 GW of Enel's capacity were split into three GenCos to be offered on the market starting from 2001: Eurogen (7008 MW), Elettrogen (5438 MW) and Interpower (2611 MW). In 2002, Edison was awarded the biggest of the three GenCos, i.e. Eurogen. The latter's fleet was composed by six thermoelectric and 49 hydroelectric plants, equivalent to approximately 10% of Italy's total installed capacity at the time. Over the following years, Edison allocated recurrent investments aimed at both the construction of new technologically advanced power stations and repowering the existing ones.

As for new plants, Edison deployed combined-cycle technology even further, gradually improving the overall energy efficiency of its fleet. In 2005, two new CCGTs were built, one in Altomonte (Calabria) with 780 MW of capacity and one in Candela (Puglia) with 400 MW, while in 2006 two new plants started operations in Torviscosa (Friuli-Venezia Giulia) with 790 MW and Piacenza (Emilia-Romagna) with 788 MW. As for repowering investments, in 2006 Edison allocated €174.2 million to this purpose and other 118 million Euros were added in 2007. As a result of all these investments, the Company's generation fleet became a reference model in terms of efficiency at national and international level, while consistently increasing its sales of electricity. For example, in 2005, the thermoelectric power station in Candela was built to be the most efficient and environmentally friendly plant in Italy and in 2007, the Simeri Crichi (Calabria) power plant (857 MW) was equipped with the most advanced technologies in terms of efficiency and emissions reduction, further enhancing competition on the wholesale electricity market.

Even during a period of slower economic growth, exacerbated by an over-supply of electricity, Edison kept on carrying out the rationalisation and optimisation of its fleet and on further investing on its most flexible and efficient thermoelectric power plants. More recently, in 2018, Edison and General Electric Power Services Business signed an agreement concerning the refurbishment of the Candela plant. The deal included both hardware and security upgrades, resulting also in decreasing costs for maintenance and operations.

Edison Research Center in Trofarello (Piedmont) was founded in 1993 with the specific aim of contributing to the Company's future strategy in the field of energy efficiency. Edison's constant research activities for innovative solutions combined with recurrent investments into its own assets have significantly improved the environmental, economic and technological performance of the Italian generation fleet.

2 The Proactive Role in the Opening of the Final Market. The Key Role in the Nascent Electricity Exchange and the Services Market

Since the Bersani Decree (1999), Edison has tried to play a determinant role in the liberalized market at both wholesale and retail level. The GenCos' disposal was meant to open the market to new entrants, thereby promoting the liberalisation process and boosting the free competition. Therefore, when Elettrogen, Eurogen and Interpower, were auctioned, Edison participated in the second auction, where Eurogen was on sale. At the time, Eurogen represented the second largest operator in the wholesale market in terms of capacity. Edison participated in the auction owning 40% of the consortium Edipower, whereas Aem Milano had 13.4%, Aem Torino and Atel had 13.3% each; the remaining shares were distributed between Unicredit, Interbanca and Royal Bank of Scotland. After winning the auction with Edipower, Edison has been making offers to clients on the free wholesale markets of gas and electricity and has been working to expand its share of downstream consumers in the power and gas sector, consolidating its position just behind the former monopolists.[3]

In order to further support the process of liberalisation, Edison joined GME (Gestore del Mercato Elettrico), i.e. the Italian power exchange, at its opening in 2004. During the first year, already 30% of the national consumption was traded on GME and Edison's sales increased by 14%, reaching 50 TWh for the first time in the Company's history. In 2005, Edison supplied 17.3% of the national demand of electricity and traded 2.6 TWh on GME. In the meantime, the Company contributed to the improvement of the market regulation by taking part in the consultations called for by GME and participated to the consultation tables organized by the Italian Regulator for the creation of a gas exchange. In the same year, Moody raised Edison's creditworthiness, praising, among other reasons, the Company's ability to manage the ever-changing context of the free market. Even in 2008, when demand for electricity dropped and prices decreased, the company was able to compensate its slightly decreasing over the counter sales through its trading activities on the power exchange, whose volumes had more than doubled.

During the years prior to the opening of the retail sector in 2008, Edison kept on consolidating its leading position in the wholesale market. Indeed, marketing choices were rewarded with an increase in electricity sales to large, small and medium-sized companies as well as wholesalers on the free market; sales on the gas market were also on the rise and Edison got a leading role on GME (Fig. 2).

[3]Relazione Annuale sullo stato dei servizi e sull'attività svolta, ARERA 2001.

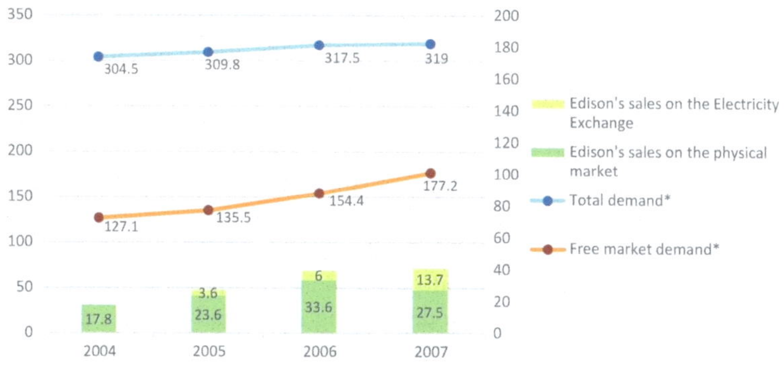

Fig. 2 Edison's over the counter sales and on the power exchange between 2004 and 2007 compared to the national electricity demand (TWh). *Source* AEEG elaborations of GRTN/TERNA data. October 2018

3 The Entry into Renewables: Wind Generation and an Innovative Investment Model

In 2018, Edison provided 7.7% of Italy's electricity consumption[4] and 19% of the renewable electricity through its 91 hydroelectric plants, 45 wind farms, nine photovoltaic fields and one biomass plant.

In particular, hydroelectric technologies have played a significant role in the Company's history since its beginning; in 2018, hydro represented 65% of the Edison's total generation from renewables. The first plant was built on the Adda river, in Lombardy, in 1898, followed, few years later, by two other plants on the same river (the first entitled to Carlo Esterle, in 1913, and the second entitled to Guido Semenza, in 1920). In 2018, the Company reported a total capacity of around 1132 MW, 70 of which distributed between 32 mini-hydro plants. This growth was made possible by recurrent investments over the years, distributed both on the development of its own fleet (strengthening already owned plants or building new ones) and on acquisitions of new capacity.

While still believing in the strategic role of hydroelectric plants, in recent years Edison has decided to focus on becoming the first national producer of wind energy. In 2004, the Company added 40 MW to its portfolio by building new wind farms in the areas of Faeto-Castiglion Messer Marino (Abruzzo) and San Bartolomeo Volturino (in Puglia) and, as a result, its wind energy production reached 404 GWh, an increase of 24% with respect to the previous year. Edison went on gradually expanding its fleet of renewable energy until 2006, when it achieved a leading position in wind, with 260 MW of installed capacity. At that moment, coherently with its core vision of a sustainable future made of renewable

[4]Indagine annuale sui settori regolati, ARERA 2018.

energy, CO_2 emissions reduction and energy efficiency solutions, Edison set an internal target to double its wind capacity within the following six years. In the same year, it opened two new wind farms in Ripabottoni (in Molise) and in Sella di Conza (in Campania), totaling 26 MW, and the production of the wind sector alone grew by 37%. In 2007, the Company invested 22 million Euros in the Lucito (in Molise) wind farm with 34 MW of capacity and in that located in Melissa-Strongoli (in Calabria) with 50 MW. As the Company was committing increasing amount of investments to renewables (€1 billion more between 2008 and 2013), in 2009 Edison allocated €94 million to wind power alone. Furthermore, even during the economic crisis, Edison did not stop developing new wind projects and, in 2010, acquired 100% of Gamesa Energia Sa adding 26 MW to its wind portfolio. In 2012, the Company confirmed its commitment to focus on the efficient integration of renewable energy and flexible gas-fired generation.

In 2014, shortly after the first signs of recovery from the economic crisis, Edison entered into an agreement with EDF Energies Nouvelles and F2i, an institutional investor with relevant experience in the Italian infrastructures sector. The deal gave rise to E2i Energie Speciali, which combined all assets from Edison Energie Speciali and some from EDF Energies Nouvelles, becoming the third operator in Italy in the renewable energy sector with a total power of 600 MW, mostly wind energy. Ideally, the agreement was meant to reinforce each one of the three operators by putting together and harnessing their respective core competences: Edison's management and optimisation of electricity production from different energy sources, EDF's operation and maintenance skills and F2i's financial experience.

In 2016, Edison included the development of a low-carbon mix among its three strategic pillars and, using the E2i platform, accomplished two operations which were crucial in achieving a leadership in the wind energy market. The company won 153 MW of on-shore wind capacity, at the auction organized by the Gestore dei Servizi Energetici (GSE), to be added to the 600 MW already owned by E2i. In the auction, E2i presented and was awarded eight projects, requiring an overall investment of 200 million Euros. In particular, E2i proposed the development of 165 MW in total: five green-field plants in Campania, Puglia, Sicily and Basilicata and three full renovations, entailing technological upgrades of operating wind farms in Abruzzo and Basilicata; these projects increased the production while reducing the installed turbines, with a clear benefit in terms of landscape. In 2017, the Company received a funding of €150 million from the European Investment Bank, meant to support these projects in particular.

As of 2018, Edison produced 17.3 TWh of wind energy and invested €134 million for the construction of new wind farms (both greenfield and full revamping) in Vaglio (Basilicata), San Giorgio La Molara (Campania), Castiglione Messer Marino and Schiavi d'Abruzzo (Abruzzo), totalling 57.7 MW of capacity.

4 The Role in Gas Supply: The LNG Terminal Bet and Long-Term Contracts

The gas market experienced a profound change after the "Letta Decree" (2000) that revolutionized the Italian gas sector: from a closed market with a single vertically integrated state-owned monopoly, i.e. Eni, to a market open to competition in production and retail. Following the implementation of such Decree, local gas utilities started to grow in size (by number of customers, volumes and revenues) and to increase their market shares through collaborations, alliances and mergers with other gas operators. The aim was to achieve larger customer bases and higher supply volumes.

Natural gas has traditionally played a significant role in Edison's strategy as it represented a key option in the energy transition of non-nuclear countries to allow decarbonisation processes, while preserving energy system security and flexibility. Indeed, the penetration of intermittent renewables in the power grids and the slow development of storage technologies requires gas to ensure energy security. In 2017, thanks to its supply contracts portfolio and international markets expertise, Edison was recognized as a leader in the gas value chain with a focus on LNG; Edison became EDF's Gas Platform by managing the Group's assets. As of end of 2018 this business area accounted for almost half of the company Ebitda and Edison was the second largest natural gas importer in Italy (see Fig. 3) with a market share of 21.8%, as well as one of the two national operators able to manage the entire gas supply chain.

This was the result of several successful strategies and operations launched by Edison in gas sector.

Today, Edison provides safe, flexible and competitive gas supply to its customers, to Italy and to the rest of Europe, thanks to four long-term contracts with some of the top producers for a total of 14.4 bcm a year (see Fig. 4). This was the result of successful renegotiations of long-term gas supply contracts with big energy exporters with the aim of reducing the previously agreed contractual prices. Thanks to these agreements Edison emerged from the difficulties in which all European gas companies with long-term take or pay commitments entered into in the last few years.

The long-term supply contract in Qatar with RasGas concerned the LNG that Edison regasified at the Adriatic LNG Terminal, located in the northern Adriatic Sea. This terminal was the first ever offshore Gravity Based Structure (GBS)[5] for unloading, storing and regasifying LNG and it had a regasification capacity of 8 bcm of natural gas per year, or approximately 10% of Italy's current natural gas requirements.

[5]A Gravity Based Structure (GBS) is a fixed concrete structure laying on the sea floor. The structure is installed with the LNG storage tanks and regasification equipment.

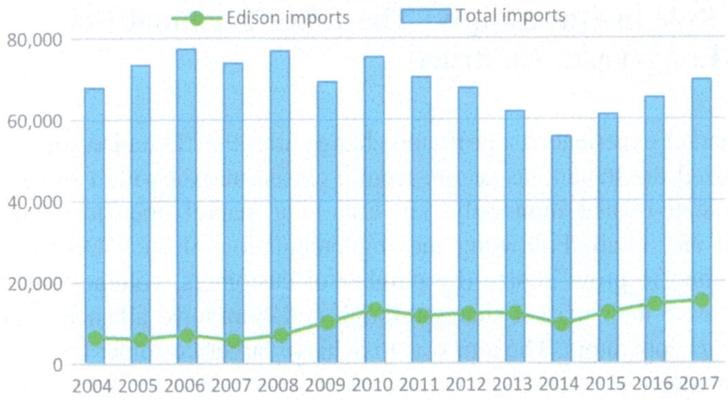

Fig. 3 Edison's natural gas imports from 2004 to 2017 (Mcm). *Source* ARERA, 2018

Exporting Country	Gas Producer	Quantity (bcm)	Expiration Year
Libya	Eni	4	2028
Algeria	Sonatrach	2	2019
Russia	Promgas	2	2019
Qatar	RasGas	6.4	2034

Fig. 4 Edison's long-term supply gas contracts

The Adriatic LNG terminal is still a strategic infrastructure for Italy as it opened a new gas route which was totally independent from existing pipelines; it represents a remarkable step towards the diversification of energy supplies. 80% of the terminal capacity was allocated to Edison for a period of 25 years.

In addition to the mentioned four long-term contracts, the Company opened a new supply channel from the United States, which had recently started to export LNG, following the "shale gas revolution". In 2017, Edison signed an agreement with Venture Global to purchase a million tons a year (for 20 years) of LNG from Calcasieu Pass plant, developed by Venture Global in Louisiana. The Company is also committed to the development of new natural gas import routes with a significant investment in the new infrastructure of the so-called Southern Corridor, which was necessary to provide a natural gas supply route from Caspian and Middle Eastern regions to Europe.

5 Back to the Future: The Latest Generation CCGT Technology for Renewables Development; the Small-Scale Gas Supply Chain to Support Sustainable Mobility

In the last few years, Edison reorganized its power generation portfolio towards a more renewable and decarbonized energy mix following the growing **international** and national attention to climate change and the challenge to urgently peak global GHG emissions. More specifically, the new plan was aligned to the Italian strategic view on energy that was encapsulated in the *Strategia Energetica Nazionale* (SEN) and in the *Piano Nazionale Integrato per l'Energia e il Clima* (PNIEC).

In order to increase its power capacity, Edison power generation plan was based on a combination of M&A, targeting specific market opportunities in the area of renewable sources, and organic growth, thanks to the development of many medium-small scale projects, with high probability of success, using proprietary technologies and developing innovative solutions.

Edison's production fleet in Italy consisted of 153 plants with an overall capacity of 6.1 GW. Around 80% of the electricity that Edison produced in 2018 was generated thermoelectrically, thanks to 14 power plants, all combined gas cycle, for an installed capacity of 4.6 GW. Today, Edison is continuing to invest in innovation, focusing on the best *Made in Italy* technologies in order to support the Country's industrial growth and sustainable development.

One pillar of the Company's strategy was the consolidation and further development of efficient, flexible and low environmental impact thermoelectric generation, to complement the renewable sources. First, Edison worked on the reliability of its combined cycle gas plants, in order to increase the performance and contribute to the safety of the national energy system. Moreover, the company started to study the development of a new generation of gas-powered plants, leveraging the specific competencies in engineering and innovation.

In particular, in 2019 Edison launched the project for upgrading the industrial site of Porto Marghera with one of the most advanced Italian technologies, developed in partnership with Ansaldo Energia, an Italian leading player in the power generation industry. When completed, the plant will be equipped with a high-efficiency 780 MW gas turbine (GT36) that will power the combined cycle of Marghera. With an energy efficiency of 63%, this will represent one of the best available technologies in Europe. Moreover, such efficiency will lead to a reduction in CO_2 and NO_x emission of about 40% and 70%, respectively, when compared to the average of the current Italian thermoelectric fleet. The new and latest-generation combined cycle of Porto Marghera is part of Edison's investments plan in Italy (over the period 2019–2022) to support the Country's energy transition, ensuring the security and flexibility of production required to balance the intermittent nature of renewable sources.

Moreover, in 2018 Edison signed an agreement with PIR Group[6] and Knutsen[7] to launch the first Small-scale LNG (liquefied natural gas plants) integrated logistics chain in Italy, constructing a coastal depot in Ravenna and acquiring a vessel for transporting LNG. This ambitious project aimed at building the first costal LNG depot system—starting with the Ravenna Costal depot—to provide gas for both maritime transport and heavy-duty vehicles, in order to abate pollutants and support the transition to a more sustainable mobility. In fact, when liquefied, natural gas has a much smaller volume and can be transported more easily along more flexible routes than gas pipeline networks, such as via the sea. Once it has reached its destination, the liquefied natural gas can be used as a fuel for heavy-duty transport or can be stored in small-scale plants.

The deposit, which will start to operate in 2021, will have a storage capacity of 20,000 cubic meters of LNG and will handle more than 1 million cubic meters of liquefied gas a year, making LNG available in Italy to fuel at least 12,000 trucks and up to 48 ferries a year. The Italian market for LNG as an alternative fuel to traditional oil products is marked by strong potential for growth subject to the development of new infrastructures. In Europe, it is forecasted that approximately 280,000 LNG trucks will circulate in 2030 while 244 ships with LNG propulsion are already operating or under construction today.[8]

6 Services and Customers: Enhancing Positioning Along the Supply Chain and Innovation to Create Value for Customers and Territories

Starting from 2008, Edison decided to enter in the downstream business, seizing the unique opportunity of the opening of the Italian retail market. Customers were positioned at the center of the new strategic plan, and the Company started a growth strategy leveraging on the fragmentation of the Italian market, through important acquisitions, focusing on local suppliers of power and gas, and energy service providers.

In this perspective, Edison closed several M&A operations in the past years with the goal to further expand the customer portfolio and create a stable relationship with clients. In 2018, the acquisition of the commercial activities of Gas Natural Vendita Italia was a fundamental step to consolidate this pillar of the new strategy. The investment has allowed to increase the customer base by 50% (500,000 customers) and to expand Edison's presence in the central-southern Italy. This operation was combined with the acquisition of Attiva Puglia (2018), a local utility

---

[6] PIR Group provides advanced logistic solutions for all needs of storage, handling and distribution of bulk liquids, dry goods and packaged products.

[7] Knutsen is a fully integrated shipping company with operation in multiple shipping segments.

[8] Edison (2018), Press Release published on November 30th, 2018.

operating in the gas retail in the South of Italy (Puglia region) with a portfolio of about 30,000 customers.

As of the end of 2018, the Company served customers through more than 1.59 million contracts with 13.7 TWh and 7.3 billion m^3 of electricity and gas sold, respectively (see Fig. 5). The downstream development strategy aims to reach a customer base of 2.5 million clients in 2022 through both organic growth and M&A operations. Moreover, the goal is to further develop a high value-added offer for the retail market and innovative energy services for its customers.

Another important area of development was the energy and environmental services solutions. Edison aimed at increasing the quality of its customer services and the satisfaction of the buyers in order to make the most out of the deregulation process that was taking place in Italy. One key example of such strategy is represented by Edison's Energy and Environmental Services division; it was founded with the specific aim of providing a wide range of energy and environmental services to industrial, commercial and retail customers. In 2018 Edison completed the acquisition of Zephyro S.p.A.,[9] a leading player in the provisioning of integrated energy management solutions for the public administration (e.g. hospitals), consolidating its market position in the segment of energy services.

Furthermore, the company in 2017 launched a new offer for the residential market, Edison World. This innovative and integrated platform provides specific product lines, and included several services targeted to families, such as self-consumption solutions, smart home technologies, maintenance and repairment services, energy efficiency management and insurance.

Always in 2019 Edison has consolidated its offer of innovative customer services by acquiring Assistenza Casa, the Italian subsidiary of the international group HomeServe, a firm with a network of over 1400 professionals operating on the territory and about 300,0000 clients. This acquisition has strengthened Edison's portfolio by enhancing its electricity and gas offer with a complete range of services dedicated to maintenance, repairment, installation, assistance and smart home.

Within this business line, Edison introduced in 2018 Plug&Go, an innovative offer to make electric mobility accessible to all. This initiative included a long-term car rental solution, the installation of the Wall Box (the physical system that provides electrical power to the vehicle plugged in by a cable) for domestic recharging and the recharge service. This strategy was targeting the growth expectation of the electric mobility market in Italy, where, despite the number of electric vehicles is

[9]Zephyro, whose shares have been traded since December 2015 on the AIM Italia (Alternative Investments Market managed by Borsa Italiana S.p.A.), was a leading operator in Italy that offered integrated energy management solutions. The Company in 2017 reported a turnover of €69 million in revenues and €15.9 million in Ebitda.

	Unit of measurement	2016	2017	2018
Total Customers *	**no.**	**1,041,115**	**1,059,316**	**1,592,427**
Electricity customers *				
Total	**no.**	**537,603**	**572,858**	**656,221**
	GWh	**11,582**	**10,928**	**13,785**
Residential	no.	446,919	433,218	484,702
	GWh	1,218	1,18	1,219
Business	no.	4,098	67,322	93,094
	GWh	9,325	8,855	11,56
Small business	no.	86,586	72,318	78,425
	GWh	1,039	893	1,005
Natural gas customers*				
Total	**no.**	**503,512**	**486,458**	**936,206**
	mil Sm3	**6,531**	**6,911**	**7,372**
Edison Energia residential natural gas customers	no.	366,288	343,348	758,104
	mil Sm3	354	349	535
Edison Energia other natural gas customers (industrial companies and non-industrial wholesalers)	no.	981	6,667	12,657
	mil Sm3	6,112	6,497	6,759

* The method of calculation of end customers was modified with respect to 2016, and, from 2017, the number of sites is considered. From 2018, the residential and industrial items of AMG Gas were merged and the row relating to Attiva was inserted, a company acquired in 2018. Data do not include Fenice customers (398 electricity customers and 21 gas customers).

Fig. 5 Edison's number of customers (Edison consolidated non-financial disclosure 2018)

marginal, the increase has been of 40% in 2017, and the figure is expected to grow in the following years.[10]

Another key aspect of the downstream business activities is the historical role played by Edison with the local communities where it was operating. The Company's approach is focused on the creation of shared value on the territory, in which the tangible and intangible assets of the Company are made available to the community to address collective requirements. An example is "Edison Crowd for Palestro" initiative, launched in 2018: with this project Edison promoted the

[10]In 2018, the global electric car fleet exceeded 5.1 million, up 2 million from the previous year and almost doubling the number of new electric car sales. The Global EV Outlook has argued that in 2030, global electric car sales are expected to reach 23 million and the stock will exceed 130 million vehicles.

development of mini-hydroelectric plants close to Pavia, Vercelli and Novara's provinces with the objective of sharing the economic advantages with the local community and generating value for the territory. This was an innovative initiative, the first one of this kind undertaken by an Italian energy company.

ERG: An Example of Swift Change of Business Towards Sustainability

Alessandro Garrone

Abstract For over 80 years ERG has been synonymous with energy, entrepreneurship and sustainability. ERG was founded in Genoa in 1938 and for 70 years operated successfully in the oil sector, contributing to Italy's energy evolution. Our transformation began over 10 years ago and has seen us become one of the leading independent electricity producers from renewable sources in Italy and Europe with a portfolio of over 3 GW comprising wind, hydroelectric, photovoltaic and high-efficiency cogenerative thermoelectric power plants. In the immediate future we want to continue contributing to the decarbonization path through a multi-year business plan completely dedicated to renewable sources of electricity in Italy and Europe, involving almost €1.7 billion of investments and an additional capacity of 850 MW within 2022. ERG has shown how it is possible to speed up the process of decarbonisation, today an integral part of the agendas of countries all over the world and particularly in Europe, as ratified by the 2015 Paris Agreement, the Clean Energy Package and the European Commission agenda. The following pages summarise our path—which chimes with that of the 20-year AGICI Utilities Monitor—but above all convey what it means to be and to work for ERG: our values, our business approach and our vision of the future of energy in terms of fight against climate change.

Keywords Energy · Renewables · Sustainability · Innovation · Repowering · #greenenERGymakers · #WeAreERG

1 ERG. 80 Years of Energy Challenges

As reminded in the recent book on company's 80th anniversary, ERG's philosophy is defined in great depth by some considerations of its founder in an old letter to his employees.[1]

[1]Taken from: Alessandro Plateroti, introduction to the book "ERG, da sempre un passo avanti", Codice Edizioni, 2018.

A. Gilardoni (ed.), *The Italian Utilities Industry*,
https://doi.org/10.1007/978-3-030-37677-2_6

> Ever since I founded the company, back in the days when we had just a few dozen employees, I have constantly strived to make the relationship between myself and my workers warmer and more human. And I believe that this business approach, to my mind essential, has produced tangible and positive results because [...] it has generated an affinity between us, one of the best characteristics of my company and a factor that has enabled us to create a real "family" whose every member is fully committed to acting in the best interests of the community. I am convinced that this characteristic and this "family" will continue to exist in the future even if the dimensions of the company should change.[2]

Italian entrepreneurial history has always been considered one of the most varied in the world and has been admired and studied at an international level. It is a story of men, women and families, of visionary entrepreneurs and proud craftsmen who, generation after generation, and with no economic system behind them, have transformed their ideas into great national and international businesses, playing a key role not only in Italy's transition from an agricultural to an industrial economy, but also in the social and economic development of the country and, above all, their local regions.

But even if every family business, just like every family of businessmen and women, has its own story which makes it unique and inimitable, there is a common denominator that down the decades and generations has represented both a fundamental value of the Italian approach to business and a development model that no multinational or public company has been able to replicate: the knowledge that the success of a business depends not only on the availability of capital and technological prowess, but above all on the deep understanding of the geographical and cultural ecosystem and the characteristics of the region, its inhabitants and the working world.

The book on the 80th anniversary of ERG is not simply a celebration of the company's history and the tenacity and intuition of its founder and his successors; it also tells the story of a model of industrial development that played a decisive role both in the post-war reconstruction of Italy and the entry of its economic system into the broader process of globalisation. In other words, the text and photos recount much more than the long and fascinating history of the Garrone family, from the first generation to the present day. They also tell the story of the early days of globalisation and the impact of the 'cultural factor' on business management and strategy.

These considerations already formed the basis for the book 'From oil to energy: ERG 1938–2008', published ten years ago to mark the company's seventieth anniversary.[3] But this new publication offers much more than a mere celebration of the past and the strategies that made ERG a major European refining and petroleum distribution company. The text and photos in this new edition express the essence of the business culture and values that have enabled the third generation of the Garrone family to reinvent the Group's mission—the definitive exit from the oil

---

[2]Taken from: Edoardo Garrone, letter to employees, 20 June 1963.

[3]"From oil to energy. ERG 1938–2008. Story and business culture" by Paride Rugafiori and Ferdinando Fasce, Laterza, 2008.

sector is now complete—and turn its attention to the new global energy model: green energy and sustainable economy. The ERG Group is now Italy's leading wind energy operator and one of the top ten on-shore wind operators in Europe.

In the coming decades ERG will no longer be the company envisaged by Edoardo Garrone in 1938 but will continue to grow and thrive from those same family and cultural roots: industrial and social responsibility and, above all, commitment and hard work. After all, ERG is not simply the initials of Edoardo Raffinerie Garrone; erg, or ergon, the Ancient Greek for "work" (ἔργον), is also the unit of measurement of both energy and work. As the saying goes, nothing happens by chance.

2 ERG and Its First 80 Years in the Energy World

Figure 1 on next page summarises the key milestones in ERG's history. The various themes are discussed on the following pages.

2.1 From the "Internal" Post-war Refinery in Genoa to the Fuel Network and Through to the Cargo Refinery of Priolo

The ERG story began officially in Genoa on 2 June 1938 although in reality its roots can be traced back to 1923. Major international businesses were already active in Genoa at this time and Edoardo Garrone acquired his first experiences managing a number of companies involved in refining mineral oils. In 1938 he decided to take the first step towards the birth of Raffinerie Edoardo Garrone, setting up a sole proprietorship which would lead to the construction of the San Quirico refinery on the right bank of the Polcevera river in the immediate hinterland of Genoa. Work began but no sooner had he celebrated the launch of this new venture that the Second World War broke out.

At the end of the 1940s the sands shifted again: oil was becoming integral to the economic recovery and Italy was beginning to acquire an increasingly strategic role as the geographical meeting point between Europe and the Arab and Middle Eastern countries. Its Mediterranean ports have deep waters suitable for the docking of oil tankers. Around the same time the ERG name began to establish itself in the still-small Italian refining industry; at the start of the 1950s ERG also decided to enter the world of distribution with around 300 petrol stations in Liguria, Piedmont and Tuscany.

The sudden death of the founder in 1963 and his immediate replacement by son Riccardo—just 27 years old—brought another of ERG's distinctive values to light: the importance of the family, which would prove to be an essential resource for the

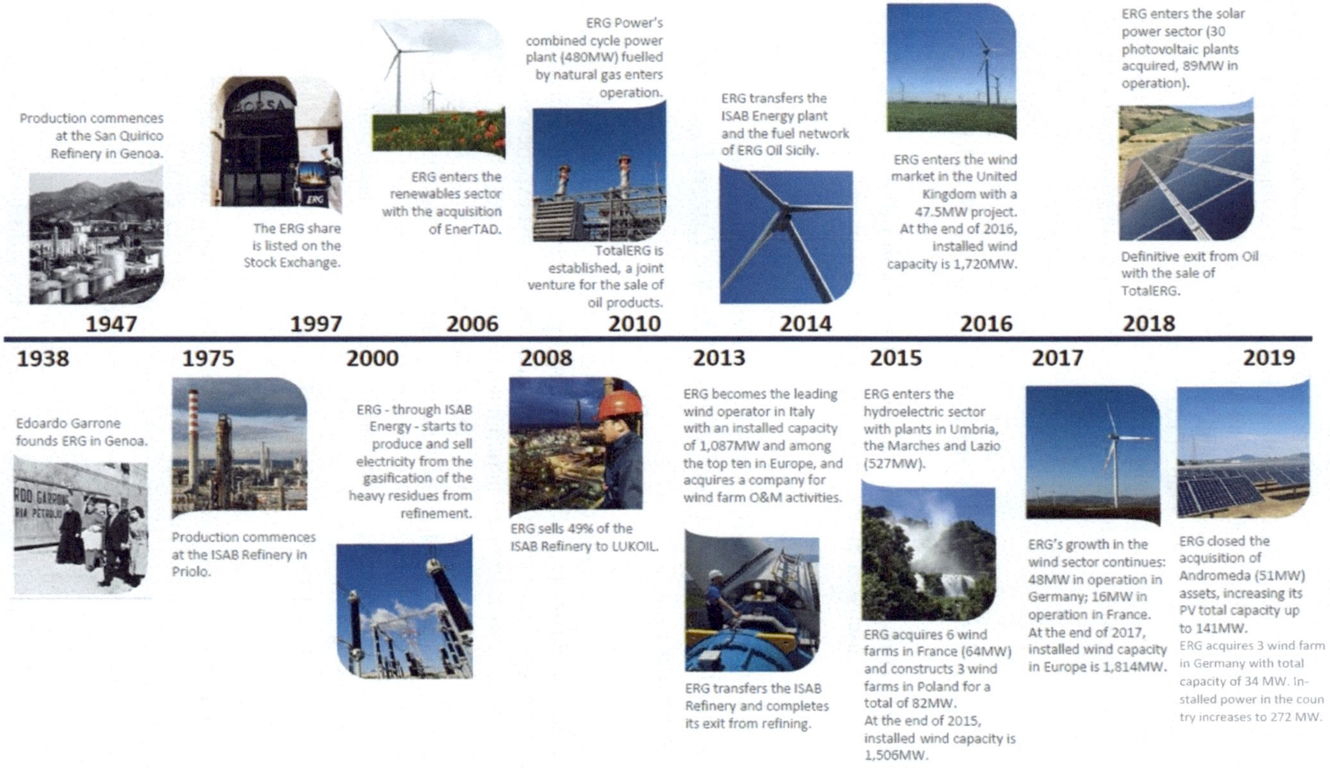

Fig. 1 Key milestones in ERG's history

continuity and development of the company. In a fast-developing context, for ERG the aim was to continue along the path marked out by the founder and this took concrete form in the extraordinary investment made to construct a petroleum logistics system consisting of 35 km of oil pipelines between Genoa and Arquata Scrivia and a depot with storage capacity of around 1 million cubic metres. An ambitious project that would create a direct link between the Genoa refinery and the rich fuel markets in the Po Valley and abroad.

Consumption forecasts were good, so much so that the San Quirico refinery risked falling behind the competition in terms of profitability and found itself increasingly "squeezed" by the surrounding city. In addition, the idea of developing new refining hubs in lower Piedmont—location of the Arquata depot—came up against insurmountable obstacles. An alternative solution was required, and an opportunity presented itself when the company received a proposal to form part of the ISAB—Industria Siciliana Asfalti e Bitumi—project for a large refinery in Sicily, at the petrochemical hub near Syracuse, which could produce low-sulphur fuels. This was well before the regulations were tightened.

The refinery, one of the biggest and most complex in the sector, was the last to be built in Italy and one of the last in Europe, and the investment was colossal. Eastern Sicily was chosen because of its strategic position along the "oil route" that connects both North African and Middle Eastern crude oil producers via the Suez Canal, and the Russians via the Bosphorus.

Everything was set up nicely for the investment with the demand curve for petroleum products on the rise. But nobody could have predicted the drastic effects that the Fourth Arab-Israeli War would have on the world energy market. The first big oil shock took place in 1973 but the construction of ISAB continued and crude oil processing began in 1975, just as consumption levels had quickly begun to plummet.

But for those in the know, moments of crisis also present some of the best opportunities. ERG knew that its strengths lay in its agility and its capacity to quickly adjust its business plans and it therefore began to work on a new path that would see it grow and establish itself on the fuel network. It understood that the only way of limiting its market risk was to sell the goods that ISAB produced.

In 1984, at a time when the major foreign oil companies were deciding to leave Italy, it once again bucked the trend by acquiring the entire Italian fuel network of French company ELF and following this up with the acquisition of Chevron Oil Italiana. Thanks to these two operations ERG controlled around 2800 petrol stations, creating an ERG network that spanned all of Italy. This new major business was called ERG Petroli. In the meantime, 12 refineries closed in Italy, including the old San Quirico refinery.

The Group acquired a controlling stake in ISAB, eventually taking it over completely in 1997. However, the volatility of the market was no secret and in the meantime ERG began plans to further diversify its portfolio. The result was a joint project with Edison Mission Energy to construct the ISAB Energy complex next to the refinery: with its 528 MW capacity, it was the first IGCC (Integrated Gasification Combined Cycle) plant in Italy and produced electricity directly from refining residues through the gasification of the heaviest oils. An unprecedented industrial operation in Italy, achieved thanks to a 1800 billion lira project financing agreement.

This plant, which began operations in 2000, produced electricity, steam and hydrogen while at the same time avoiding the production of millions of tonnes of high sulphur-content fuel oil for consumption every year. ERG was therefore the pioneer of a different and more modern concept of the refinery, now seen as an integrated energy centre rather than a plant for the transformation of crude oil. The launch of the ISAB Energy IGCC plant marked ERG's entry into the electricity sector, hailing the beginning of a multi-energy strategy that sought to increasingly move the Group's focus from refining to the production of electricity with the aim of reducing its reliance on the intrinsically volatile oil industry.

Unlike many traditional family-run businesses, over time ERG also decided to entrust its management with the running of the company, making transparency and efficiency its distinctive traits ahead of its listing on the stock exchange. In October 1997 ERG's shares were listed on the Italian electronic stock market for the first time and the operation was a major success with demand ten times greater than the share offering. The Group's flotation on the stock market gave it the necessary size and strength to contain market risks, consolidate and, most importantly, exploit the diversification opportunities generated by the deregulation of the energy sector. In fact, at the end of the 1990s the world and European energy production panorama underwent a period of structural and regulatory transformations connected on the one hand with the growing sensitivity to environmental, climatic and geopolitical issues, and on the other to the freeing up of the energy market.

In the early 2000s—a period of relative stability for the oil industry and an important time of consolidation for ERG—nobody could have predicted the oil crisis and the subsequent financial crisis that would take hold at the end of the decade. Having navigated the ever-changing international oil panorama for 70 years, the Group had come to the conclusion that the time was right for a change. In fact, whereas refining technology had once largely been the domain of the West, knowhow had now spread to various countries in the world, some of which were undergoing major economic development. Many new refineries were therefore built directly in the crude oil-producing countries and consumption began to shift eastwards: the combination of the latest available technologies and the reduction of crude oil transport costs made these refineries more efficient, profitable and better positioned than the existing ones. This was one of the effects of globalisation, which reduced distances, increased geopolitical instability (particularly in the Middle East), shook up the banking industry (with the subprime mortgages crisis) and saw many countries enter into recession for the first time since the end of the Second World War.

Well before the emergence of this scenario ERG realised that remaining anchored to the downstream oil market was too risky unless it had its own oil wells. Furthermore, although they were perfectly efficient the ISAB refinery plants were beginning to suffer the competition of the more modern *supersites* outside of

Europe. A period of reconnaissance therefore began with the aim of identifying possible oil business partners from among the global industry leaders.

In these same years Russian giant LUKOIL displayed its interest in the ISAB refinery. One of the biggest oil companies in the world, it was now interested in penetrating Western Europe and had identified Sicily as an initial logistics hub from which to develop this strategy. Negotiations progressed immediately and in 2008 a joint venture agreement was signed whereby ERG retained 51% and sold 49% to the Russian colossus including a put option that guaranteed ERG the right to withdraw if the joint management of the refinery proved too difficult.

In terms of the macroeconomic climate, at the end of 2008 the price of crude oil fell as, more importantly, did that of petroleum products: the operating margins of the refinery began to slip deeply into the red and for ERG, which didn't have the "broad shoulders" of LUKOIL, covering these losses in the long term was not sustainable. Faced with this evidence, it soon became clear that the only way out was to leave the refining business completely. The Group therefore began the process of permanently exiting the oil sector, selling its shareholding in the refinery to LUKOIL over a period of three years.

2.2 *Towards Electricity Generation*

The deregulation of electricity production and the wholesale energy market are the cornerstones of ERG's transformation from leading private Italian oil group to top player in the production of electricity from renewable sources and high-yield gas cogeneration. However, it was thanks to the deregulation permitted by the Bersani Decree[4] that entire sectors, traditionally state-run, were able to free their potential and open up to private enterprise via an increasingly free and competitive market.

Driven by the deregulation of the electricity sector but also the spread of environmental awareness, with the themes of climate change and global warming of increasing importance in the energy world, ERG began a new transformation which in the space of just a few years would radically change its market position. In fact, ERG used the liquidity deriving from the sale of its oil assets to make major investments in the production of electricity from renewable sources and high-efficiency plants, carrying out over €8 billion of divestments and acquisitions in around ten years.

ERG didn't stand around watching when it came to thermoelectric power either: the extremely busy industrial site of Priolo (SR) was home to various businesses that produced interdependently; one of the biggest of these was the ISAB refinery

[4]Published on 16 March 1999, the Bersani Decree was the regulation act that transposed EC Directive 96/92/EC of the European Parliament and Council of 19 December 1996 into Italian law. This Directive introduced common regulations for the generation, transmission and distribution of electricity with the goal of creating a domestic energy market and opening to the market where it had not yet been liberalised.

but there were also the petrochemical plants and the ancillary services for the factories. The activities of the ERG thermoelectric power plants located on the site of the ISAB refinery were closely integrated with these plants and the steam, electricity and demineralised water they produced were key to their functioning. ERG decided to replace these old plants with a modern high-yield natural gas-powered Combined Cycle Gas Turbine (CCGT) which became operational between 2009 and 2010 and would become one of the cornerstones of the Group's energy mix.

The electricity produced by the 480 MW capacity plant has a dual destination: the majority is injected into the national electricity grid while the rest is used to power the industrial plants of the petrochemical site via a 150 kV private network; the plant respects the most stringent environmental criteria and enables reductions of 60–70% in emissions of the main pollutants. It was also the first and the largest plant to be awarded high-yield cogeneration status by the GSE, entitling it to white certificates, the efficiency documents awarded to plants that achieve major energy savings.

2.3 *The Beginning of the Metamorphosis: From Domestic Oil Company to European IPP*

In 2006 the Group developed the wind energy business with the acquisition of EnerTAD, a listed company with a portfolio of around 200 MW of wind energy distributed among its projects and operational wind farms. This period also saw the start of an international expansion process with ERG developing its presence in six other European countries—France, Germany, the UK, Poland, Romania and Bulgaria—and establishing itself as one of the top ten operators in Europe.

2007 was the year of the first cross-border operation with the acquisition of five wind farms in France for a total of 55 MW. In Italy, after the Sicilian wind farm of Vicari (37.5 MW) in 2008, the plants of Fossa del Lupo (97.5 MW) in Calabria and Ginestra (42 MW) in Campania were built in 2011, the Amàroni plant (22.5 MW) was developed in 2012 and the Palazzo San Gervasio (34 MW) wind farm in Basilicata was constructed in 2013. The ERG portfolio was also expanded in 2011 with five wind farms located in the provinces of Benevento and Avellino, for a total of 112 MW. To foster development in Eastern Europe, the joint venture LUKERG Renew was created with LUKOIL; a few months later the company expanded its presence to Bulgaria, with the purchase of an active 40 MW wind farm, and then to Romania with the construction of an 84 MW farm in Tulcea.

However, in December 2013 ERG's wind energy business really moved to the next level when it announced its acquisition of IP Maestrale from Gaz de France: 636 MW of installed wind power, of which 550 MW in Italy and 86 MW in Germany. Through this operation, described by Bloomberg as the biggest M&A deal of the year in the global green energy sector, ERG immediately doubled its overall wind power capacity to 1232 MW of installed power. With capacity of over 1 GW in Italy, it became the country's leading wind energy operator.

In the meantime, in June 2009 the package of measures contained in Directive 2009/29/EC, known as the European 20-20-20 Targets, was implemented at European level with the aim of reducing greenhouse gas emissions by 20%, increasing the amount of energy produced from renewable sources to 20%, and improving energy efficiency by 20% by the year 2020.

For ERG 2013 was another important year: in October it signed an agreement for the sale of its final stake in ISAB and, straight after, another one for the sale of ISAB Energy, thus finalising its exit from the coastal refining sector. In the meantime, it continued to grow in the Eastern European wind energy market with the acquisition of two active wind farms, Gebelesis (70 MW) in Romania and Hrabrovo (14 MW) in Bulgaria.

In 2014 and 2015 ERG added a new country—Poland—to its portfolio with the acquisition of three wind farms for a total installed capacity of 80 MW. The same year ERG acquired a company to oversee the running and maintenance of the farms with the aim of completing the skill set required to manage wind energy business operating activities in a direct and integrated way, taking advantage of its industrial DNA, and of achieving significant benefits in terms of both efficiency and performance. In fact, the gradual expiry of wind energy incentives would impose even more limitations on operators with only those possessing the necessary skills, structures and agile business organisation models able to meet the economic conditions required to compete on the market.

2.4 Growth of Wind Energy Business in Germany, France and the UK

In 2015, year of the COP21 with the Paris Agreement to contain climate change, ERG accelerated its growth in France. In June its wind energy capacity in the country rose to 128 MW, while in October it further enriched its portfolio with another 11 wind farms totalling 124 MW. In the same period, six farms with overall capacity of 82 MW were acquired in Germany. The Group also acquired two companies, one French and the other German, that provide technical, operational and commercial assistance via a team of 28 professionals. ERG's presence in France, which with the new projects now equates to 360 MW, and Germany, where it exceeded 270 MW in 2019, is increasingly important and strategic for the Group and its overseas development.

ERG has been active on the other side of the English Channel since 2016, acquiring the project for the construction of the 47.5 MW Brockaghboy wind farm in the county of Londonderry in Northern Ireland (later sold) and launching other new projects in Scotland and Northern Ireland, for the first time on a merchant basis (i.e. with the absence of support mechanisms).

2.5 *ERG Acquires Terni Hydroelectric Hub*

Having made a commitment to renewable energies, the time had come to decide how to integrate the business portfolio in order to balance programmable and non-programmable sources and guarantee the continuity of production and therefore profits. While these considerations were taking place, in 2014 German group E. On decided to leave Italy, putting the Terni Hydroelectric Complex with its overall capacity of 527 MW up for sale. A long and complex negotiation process then ensued before an agreement was reached at the end of 2015.

It is one of the best hydro plants in Europe, consisting of 16 production sites in Umbria, Lazio and the Marche which use the water of the River Tiber and its high tributaries; it also includes the "Marmore" waterfall, one of the tallest in Europe at 165 metres high. A giant complex, much of which was constructed dozens of metres underground in order to protect it from bombing and therefore interruptions in the supply of the steelworks, whose wartime production was regarded as highly strategic.

The hydroelectric plant, whose production can be programmed, optimises ERG's energy portfolio because it acts as a counterweight to its other intermittent renewable energies which depend on the presence of the wind and sun. Studying the imposing hydrogeological network planned at the start of last century, ERG discovered new possibilities, including that of exploiting the water releases necessary to keep water courses alive downstream of barriers. A long-neglected source of potential energy that ERG would use by constructing mini hydroelectric plants, with capacities of between 50 and 250 kW, close to water release points.

The acquisition also included a bidding centre specialised in the complex planning of water courses to optimise sales of the energy produced on the Electricity market. Within a few months a large single control centre was created in Terni, a new energy management hub which, operating on three platforms, optimises the Group's hydroelectric, thermoelectric and wind power generation on a daily basis.

2.6 *Entry into PV Segment, Strengthening of Wind Business Abroad and Complete Exit from Oil Industry*

However, one final piece was still missing from the ERG portfolio and in 2017 work began on another key operation that would further consolidate its position in the renewable energies market. This last step was its entry in the photovoltaic sector through the acquisition of ForVei, ninth biggest PV operator in Italy, owner and manager of 30 plants for a total of 89 MW of installed power. This first acquisition in the photovoltaic sector was followed by another between 2018 and 2019 which increased ERG's installed capacity in the photovoltaic sector to 141 MW distributed across nine regions of Italy. With its entrance in the solar energy market,

ERG has become even greener and technologically diverse, in line with its growth strategy.

The same year ERG's electricity generation capacity exceeded 3 GW for the first time. It is quite obvious that ERG's growth path has significantly pre-empted the evolution that the entire energy system is undergoing. The company has taken the firm decision to focus on the renewable energies sector, adopting the business approach with which it is synonymous. A transformation process which as well as being unique in its sector has also been a benchmark for change in general. A business case that has intrigued and caught the attention of the media and business strategy experts for the vision shown by the management, the level of execution demonstrated by the company and the flexibility of people in adapting to the evolution of the business model.

3 ERG's Future Has Begun: The 2018–2022 Business Plan

In 2017, following the brilliant results for the year and having achieved the set objectives of the previous business plan early, the management decided to present the 2018–2022 Business Plan one year ahead.

The plan envisages 1.7 billion euro of investments, far more than the 50% stock market capitalisation, and has three directives: the consistent development of wind energy abroad through greenfield and co-development projects, technological renewal through the repowering and reblading of the Italian wind fleet, and growth through M&A operations aimed at accelerating development and optimising the generation portfolio. 92% of investments into the business plan is destined for development, with the goal of increasing installed power by about 850 MW, bringing it up to 3600 MW by 2022.

3.1 Further Growth in Wind Energy Abroad

For growth abroad, ERG is looking to the countries where the regulatory aspect is more consolidated and reliable, like France and Germany, or where there are exceptional wind resources available, like in the United Kingdom. In line with its Business Plan, in 2018 ERG acquired four operating wind farms in France for a total capacity of 42 MW, as well as a pipeline of projects for 750 MW in total. The following year six more wind farms were bought, bringing French installed wind power capacity up to 360 MW. In Germany, ERG built a new 22 MW wind farm in 2018, winning one of the onshore wind auctions, which was followed by the purchase of three wind farms for a further 34 MW, bringing German installed wind power up to 272 MW, and accompanied by a pipeline of wind projects for a total capacity of 224 MW. In the United Kingdom, ERG purchased an authorised project in 2018 to create a wind farm in Scotland with a capacity of about 80 MW, set to be operational by March 2022. In Scotland and Northern Ireland there are also two other 49 and 47 MW development projects under construction, together with another 25 MW project in Northern Ireland to be operational by April 2021.

3.2 *The Repowering Plan for the Existing Fleet*

For Italy, ERG's Business Plan envisages the technological upgrading of a considerable part of its wind farms. Thanks to its solid business know-how, the Group in fact intends to repower the technologically obsolete and small wind-power generators. It is a major industrial project with great environmental advantages, allowing the life span of wind farms to be extended, increasing their production fourfold, halving the number of turbines and avoiding to occupy additional land. The existing infrastructures, like the electrotechnical equipment, cable ducts and access roads, can, for the most part, be reused.

Repowering wind farms, as underlined both by the National Integrated Energy and Climate Plan as well as by the recent recast of the EU Renewable Energy Directive is indispensable to Italy to reach the 2030 energy-climate goals in terms of increasing the production of electricity from renewable sources. It also brings advantages from an environmental and scenic viewpoint, since reducing the number of wind turbines diminishes the so-called "forest effect" while at the same time producing more "green" electricity.

The other kind of technological upgrading on which ERG is focusing is reblading, which consists of the replacement of old wind blades with new generation longer ones that are far more efficient: the power of the wind-power generators remains the same, but production can increase by up to 16%, again in this case without affecting the area (Fig. 2).

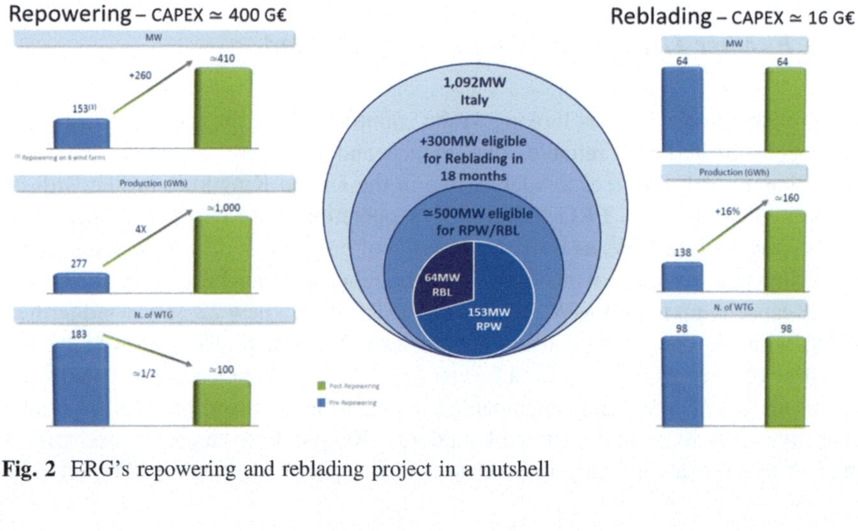

Fig. 2 ERG's repowering and reblading project in a nutshell

3.3 *Wind Repowering in Italy: A Big Opportunity*

Repowering initiatives, despite their clear benefits and EU support (RED II directive), in Italy have the same lengthy authorisation procedure as green-field plants (which takes 5.5 years on average) and—the only country in Europe where such a restriction exists—they cannot access state auctions.[5]

To highlight the importance of repowering in order to reach the 2030 renewable diffusion goals, ERG conducted a specific study together with a leading energy consultancy firm on the potential offered by such technology nationwide. The conclusions of the study show, very briefly, that:

- about 5.1 GW of existing wind farms, equal to about half of the total installed wind capacity in Italy, have technological characteristics and current performance levels such as to benefit from Repowering operations;
- through Repowering, the capacity of wind farms can increase between now and 2030 by 3.4 GW to reach 8.5 GW;
- thanks to the increase in power and performance due to new and improved turbines as well as to the windiness of the sites where they are installed (the first to be used for wind generation), the additional production by 2030 will be 12.1 TWh, equal to over 57% of the increase in wind energy production envisaged by the National Integrated Energy and Climate Plan (PNIEC);
- the increase in production on the same occupied land avoids the use of new land and allows for an on average 2.5-fold increase in the energy density of sites, thus dealing with one of the main concerns of the PNIEC regarding the development of renewable sources;
- Repowering allows for an over 50% reduction in the number of turbines installed, thus avoiding the "forest" effect.
- it could also activate investments for around 8.2 billion in the decade 2021–2030, other than economic benefits in terms of reduction of electricity cost, added value for the Italian industry and tax revenues.

[5]The current Italian legislation (see art. 1, c. 3, Law Decree 23/12/2013, n. 145 and Ministerial Decree 6/11/2014) prevents the vast majority of wind repowering projects from benefitting from any incentive or support scheme.

4 Corporate Social Responsibility, Sustainability, Relations with the Local Area, Shared Value and People Focus in Dealing with the Change of Business

The incredible growth and change witnessed by the global economy in the second half of the Twentieth century started the concept whereby a company's value is not just its figures, but also its ability to produce value in the medium and long term, and to be resilient and adaptable to the rapid changes of the modern world.

Sustainability and a new ethically responsible corporate culture are the principles that form the base of the concept of Corporate Social Responsibility (CSR), which began to take hold in the Eighties. The three major areas of CSR are the creation of sustainable value, respect for the environment, and attention to relations with stakeholders. With the birth of Corporate Social Responsibility those values officially became part of companies' strategic vision, driving them towards a deeper relationship with the local area and the development of activities aimed at improving the social fabric.

They are values that fully correspond with ERG's since its foundation: ERG's very own evolution has shown how it is possible to transform one's business in total harmony with the principles of sustainability and system responsibility, achieving satisfactory financial results for shareholders, increasing employment levels and developing a virtuous relationship with local areas. It is therefore no coincidence that ERG's sustainability goals, defined in accordance with the Business Plan, see us particularly committed to the development of electrical energy production from renewable sources (UN SDG 7) and to the environment (UN SDG 6 and 15), working conditions (UN SDG 8) and welfare (UN SDG 4).[6]

ERG's commitment to sustainable development is being more and more frequently recognised by different rating agencies, which constantly put it in the top quartiles of their rankings. More recently, Gaia Rating of the Ethifinance group confirmed with a score of 78 points out of 100 for the year 2019 the ever improving upward trend already registered in ratings for 2016 and 2017. ERG was put at well above average by the panel that lists the best 230 companies subject to both overall rating, and rating in their own macrosector (energy), microsector (renewables) and turnover category (>500 million Euro), as well as in each of the specific evaluation areas (Governance, Social, Environment, External stakeholders).

ERG was also included in the "Ethibel Excellence Investment Register" for the third year running and was even selected for the prestigious "Ethibel Pioneer Register", which lists, respectively, companies with above average CSR performance

[6]In 2018 we approved the new version of the Sustainability Policy and the Human Rights Policy, two important documents for guiding the Group's activities, following a business approach not limited to compliance with the laws in force in the countries we operate in, but proactively oriented to protect the environment, health, safety, communities and human rights, according to the founding principles of moral integrity, personal honesty, fairness and transparency in relations.

in their sector and leading companies in ESG actions (Environmental, Social and Governance) in their industry. In just 2019, ERG also achieved:

- B Rating in the Climate Change programme organised by Carbon Disclosure Project (CDP)
- the title of member of the ECPI Global Clean Energy Index
- A Rating by MSCI ESG Research
- 16th place, first of the Italian companies, in the "Corporate Knights Global 100 Most Sustainable Corporations in the World Index".

4.1 A Strong Relationship with the Local Area and Shared Value

A relationship between the company and its local area is essential to a productive performance and the development of the area. That's why ERG's relations are marked by its support for the area which goes beyond the mere use of local personnel and suppliers by seeking to bring added value also through other projects in favour of local communities.

For several years ERG has been promoting specific activities to support the areas it operates in, through training and education initiatives for young people, support for start-ups in the energy world and the promotion of major cultural events, like for example the classical plays in Syracuse staged by the National Ancient Drama Institute (Istituto Nazionale di Dramma Antico—INDA) and Umbria Jazz. In the past year alone, ERG's initiatives for the new generations in the areas it operates in have involved over 10,000 students.

4.2 Professional Growth of Personnel

ERG's sustainable way of operating is also seen in the attention it pays to its people, considered the company's most important asset. Personal and professional growth within the Group and constantly updating skills is an investment into the company's future: over 30,000 h of training provided each year—equal to around 6.6 days/person—is a clear testament. It is in fact thanks to people's work, their commitment and their preparation that ERG really stands out and can achieve its business goals. Training is tailored and adapted to the corporate language. They are more training programmes than courses, which accompany the person along both an individual as well as a group pathway and provide the tools for putting what is learned in the classroom into practice.

4.3 Welfare and Wellness

The attention to our people also materialises in a series of welfare initiatives that revolve around a personal health rationale in the broadest sense and in support of SDG 3 “good health and *well*-being”. Within the Group, a series of initiatives have been launched to enable people to take better care of themselves: first of all, all employees for a number of years have been offered a voluntary preventive health programme that involves clinical tests and scans to check for some of the most common cancers. In tandem, there has been a focus on “health thanks to exercise”: at the Genoa site a gym has been opened with equipment, machines for all kinds of exercise and divided into three areas—cardio, floor exercises, machines—with instructors for courses and personal trainers to guide all those doing exercises. Agreements with fitness centers have been signed at all company headquarters that will benefit all employees. Lastly, groups of cycling enthusiasts have been formed, and runners taking part in numerous and even international events in the colours of our Group.

4.4 ERG and Green Finance

Even the financial world recognises ERG’s commitment to sustainability and particularly to the battle against climate change. Between the end of 2018 and 2019, ERG launched two financial instruments linked to SDG7 “clean energy”, introducing a bonus mechanism associated with avoided CO_2 emissions based on production from renewable sources generated during the Plan. In the month of November 2018, ERG Group, one of the first in the sector, gained access to two ESG loans (Environmental, Social and Governance) for 240 million Euro. They are innovative instruments, typical of green finance, which involve measuring the sustainability and ethical impact of investments, while in April 2019 ERG issued its first Green Bond worth 500 million Euro, receiving requests worth six times the amount offered. The market’s enthusiasm once again represents investors’ major confidence in ERG, and confirms how our business model, strongly oriented towards sustainable development, is recognised and appreciated also by the financial market.

4.5 ERG and Circular Economy: Corbara Driftwood

The management of Corbara lake presented a problem: the driftwood transported by the river Tiber particularly during high waters and which accumulates on the shores of Corbara lake was defined as waste by law and dealt with accordingly. Aware of both the industrial as well as environmental importance of collecting the wood (contributing to keeping lakes and rivers, shores and banks clean, with hygienic and hydraulic benefits and minimising hydrogeological risks), the company thought that such material could be used in a way more compatible with the protection of the environment (instead of treating it as mere waste). In collaboration with the University of Perugia, ERG thus conducted an analysis for the "Identification of an integrated management system for the wooden biomass accumulated in Corbara lake", in order to assess the technical and legal feasibility of using it, for example, for energy recovery in biomass stations following the chemical and physical analysis of the material.

As of late 2018, the Corbara driftwood can be managed in accordance with the Deliberation of the Regional Council with regard to the "*Guidelines for managing plant residues deriving from green area maintenance, as well as driftwood left along the shores and banks of lakes and waterways*". These guidelines in particular provide for the option of considering the wood that accumulates on the shores of the lake as reusable material, which can be destined for activities like energy recovery, the timber industry or for the production of soil conditioner. All this in favour of a circular economy that optimises the use of natural resources and minimises the impoverishment of the area.

This is the very spirit and attention with which ERG is preparing for one of the great new sustainability challenges in the wind sector in Europe, with repowering projects, reusing replaced parts of wind turbines (mainly the blades).

4.6 *The ECO ERG Project*—Plastic and Paper Free

In 2018 ERG launched the ECO ERG project which consisted of a set of initiatives aimed at minimising the impact of the Group's activities on the environment, improving separate waste collection and recycling and spreading some virtuous practices already applied in some of the Group's sites. The project involves some different areas:

(a) ***Plastic Free***: elimination of disposable cups, half-litre bottles used by staff (about 200,000/year), water bottles from the conference rooms and for coffee breaks, plastic snack packaging from the vending machines. Each employee received a reusable thermos/steel bottle; suppliers' contracts have been reviewed so that glass bottles/jugs will be used in the conference rooms and to eliminate, insofar as possible, plastic food packaging. The project also envisages the use of low environmental impact soap, without micro plastics (extending the policy in Genoa to other sites)
(b) **Recycling**: use of an office bin for just paper, while plastic, tin and unsorted waste will be collected in the communal bins on the floor
(c) **Paper free**: highlighting the potential of office applications that avoid printing documents is being developed.

5 ERG's Long-Term Vision of Energy: Decarbonisation, Sustainability and Electrification

The energy sector has witnessed a profound change, comparable to the oil boom in the first half of the 1900s. But there is a new element, one that never occurred in the "Age of Extremes": the discontinuity that triggered this change is not linked to an economic, industrial or military need, but rather to the cross-cutting and to some extent less tangible need for anthropic initiatives to be environmentally sustainable. Global awareness of the limited available resources and the need to evolve towards development models compatible with safeguarding our eco system, plus the scientific link between the emissions of climate-altering gases of anthropic origin, the increase in Earth average temperatures and current climate changes—signed four years ago in the Paris Agreement—has spread the concept of responsibility in economic initiatives beyond the merely financial sphere to embrace the environmental and social sphere too.

In this system context, the energy sector must play a fundamental role, being one of the main culprits for current emissions of anthropic gases and currently mainly based on the use of primary sources—fossil fuels—finite and non-renewable (apart from in the extreme long term). It therefore has the obligation to make profound changes while still guaranteeing support for progress and innovation, leaving behind the use of fossil fuels in a process that can be briefly defined as decarbonisation.

Decarbonising energy thus means replacing fossil energy sources with renewable ones—particularly wind, sun and water—but also limiting their consumption by exploiting technological progress and innovation in the efficiency field. It also means facilitating access to energy, in its most versatile and "clean" forms, like electrical energy produced from renewable sources, to populations that are lacking it today, and focusing on a more widespread wellbeing, better distributed throughout the planet.

Within the framework of this momentous challenge, through the 2015 Paris Agreement, nations and institutions have taken on medium and long term quantitative commitments for mid this century, accompanied by intermediate targets that mean the process is gradual and proportionate.

It is difficult today to predict whether decarbonisation goals will be reached in the set ways and times; what does seem unequivocal however is that the process of transition for the energy sector has already begun globally and has already reached —save for some serious and unforeseeable discontinuity—its point of no return. In short, it is a question of predicting *when* energy decarbonisation will occur, because any doubts about whether or not it will already seem vanquished.

WindEurope, the European wind energy association, commenting on the recent proposals of NECP (National Energy and Climate Plan) sent to Brussels by Member States, found that a pledge is not a plan; Italy cannot escape the same criticism, since no implementation programmes and sufficiently detailed and inclusive tools have yet been defined to support the set goals and stand as a real intervention plan. We can reasonably imagine that in Italy in 2030 a significant share of gross energy consumption will come from renewable sources. Fossil energy sources will have to be increasingly held responsible for their impact on human health and on environmental damage so as to facilitate decarbonisation processes.

In this regard, one of the main instruments should be the European Union Emissions trading system (UE ETS), recently reviewed, which should be further strengthened, providing a minimum price for CO_2 emissions (the floor price) which increases over time so as to guarantee congruency with the system cost of the emission. A process which, in order to be fully effective, should lead to the progressive environmentally sustainable reform of the entire system.

At the beginning of the 2030s, the electrical sector should be marked by a generation mix satisfied more than mainly by renewable sources (hydroelectric, photovoltaic, wind, geothermal, bioenergies and other minor sources) and, to a lesser extent, by natural gas. The electrification of energy consumption will be one of the main instruments to promote energy decarbonisation, as moreover shown by the crucial role in Italy played by bringing Renewable Energy goals forward to 2020. This evolution, if managed with care, with bring a radical improvement in terms of safe energy sources and competitiveness, since it will involve a huge savings due to fuels not being sourced from abroad.

With the increase in the use of renewable energy sources for energy production, Italy will be less vulnerable to the geopolitical fluctuations mainly associated with countries that produce fossil energy sources. A significant contribution to this process will come from the development of electrical mobility, which particularly in light transport may be able to partially compensate for the transport sector hysteresis caused by the delayed construction of strategic rail infrastructures (which will progressively remove most heavy road transport). The electric car, which will use energy deriving from the new generation of wind and solar sources in the abovementioned electrification framework, will drastically reduce local CO_2 emissions compared to traditional vehicles both in terms of the quality of the energy consumed as well as the much higher yields. In line with current global trends, renewable sources will count for the vast majority of investments into the energy sector. The main sources investments will be concentrated on will be photovoltaic and wind, mainly onshore.

The success of the decarbonisation transition strategy for energy requires some basic system requisites which we can summarise as follows:

- the definition of a bipartisan political supervisory guidance, to guarantee stable energy policies and a stable legislative and institutional framework, disconnecting long term strategies from short term needs, accompanied by a plan agreed upon between the State and Regions about the use of the area and authorisations methods;
- the simplification and acceleration of the permit process, which represents the "minimum common denominator" both for Renewable Energy plants as well as for the infrastructural upgrading of the networks and the storage/flexibility systems;

- the upgrading of the electrical network, in terms of transmission and distribution. The electrical network has certain weaknesses in terms of the target Renewable Energy increases, both in Italy as well as in other respected countries;
- the chance for Renewable Energy plants to provide network services, through a more segmented market of dispatch services capable of receiving proper remuneration.

The electrical market will have to be heavily reviewed to adopt a model suited to Renewable Energy Sources—which is marked by costs that are all concentrated on the construction and maintenance phase and on negligible variable costs—by gradually reducing recourse to the current spot market which has been designed for the opposite generation mix (concentrated production and high variable cost, distributed consumption and independent from generation). It is therefore necessary to reduce liquidity at the *Mercato del Giorno Prima* (MGP) in favour of long-term sales contracts first and foremost through competitive auctions and, secondly, Corporate PPAs when they become sufficiently liquid.

The growth in renewable sources will have to be accompanied by revenue stabilisation tools (CfDs for example = *Contract for Difference*), usable in the long term. Instruments open to repowering projects and with distinct contingents for technology, thus maximising the synergy of wind and solar generation (morning, evening and night for wind, daylight for solar; summer for PV, winter for wind).

On a systems level, the progress of e-RES installations and more generally the development of the green economy, will allow for an increase in national employment both temporary (linked to the realisation phases of the plants and the machinery used there) as well as stable. As found by the Italian PNIEC, such an increase will more than offset the negative impact of the phase-out of carbon and obsolete or not sufficiently efficient thermoelectrical plants.

As regards generation connected to distribution networks, the digitalisation of electrical energy distribution and the recourse to the demand-side response will help to integrate the small and medium sized photovoltaic plants, upon the upgrading of the relative electrical control. Any delays in debottlenecking the network will be partially offset by the integration of the e-RESs into the dispatch market, thanks also to the spread of storage systems. The spread of storage systems to support Renewable Energy electrical production plants according to the set time intervals (electrochemical, pumping, power to gas, power to hydrogen), will have to be accompanied by a change in the regulatory and authorisation framework, particularly for hydroelectrical pumping stations.

Natural gas will retain an important role in electrical generation for the first years, to then very slowly decline under the pressure of decarbonisation policies and competition from storage systems. The latter, in its electrochemical version, will be industrialised at sustainable prices for mass use more quickly than we imagine today, thanks also to the development of the electric car.

Within this challenging scenario, and with a constantly evolving perspective, ERG will keep on operating with enthusiasm and making its contribution, as a conscious main player in sustainable development, to the decarbonisation process whose urgency we are all aware of today.

Part III
Utilities and Territories: Aggregation Processes, Circular Economy and Sustainability

A2A: A National Leader as a Result of Aggregations

Giovanni Valotti

Abstract The role of a modern multi-utility company is enhancing the quality of life and safeguarding health. The new technological solutions can generate valuable services for citizens and for the territories in which local utilities operate, determining new business models capable of supporting the transformation that will take place. A2A is growing with the so-called "multi-utility of territories" model with industrial operations, in order to exploit growth potential and synergies. The Group's territorial vocation is not disconnected from global trends and dynamics: global warming affects the electricity production and distribution networks by requiring higher investments to support the necessary transition process. The strategic view of the Group is designed around three macro-trends: Circular Economy, Energy Transition and smart solutions. The uniqueness of A2A multi-business structure is the opportunity to address these three trends with fully industrial solutions on an urban, regional and national scale.

Keywords Multi-utility · Distribution networks · Territories · Urban areas

1 Birth of A2A, Within the Evolution of the Utilities Market in Italy

The landscape of Italian utilities controlled by public entities is vast, with about 8,000 municipalized operations of varying dimensions. There are listed companies with over €1 billion Ebitda and companies with revenues of a few thousand euros, a single shareholder and a single customer. It is a galaxy of companies born also thanks to in-house contracts, which, for some years now, have been dealing more and more with the market. In order to narrow the field of analysis, we refer in particular to utilities operating in the following businesses: sale and distribution of energy, gas and heat, waste collection and management, water cycle. The considerations can also be extended to other public local services, such as public transport, a crucial element in the pursuit of the development of internal demand, the protection of the environment and the protection of health. From the beginning of the

A. Gilardoni (ed.), *The Italian Utilities Industry*,
https://doi.org/10.1007/978-3-030-37677-2_7

1900s, the development of the utilities was rapid, particularly in central-northern Italy, following a logic of reliance on in-house companies for the management of services within the competence of local authorities, and with almost no competitive pressure. Until 2002, the utilities sector was characterized by the presence of numerous medium-small sized operators, active mainly on a municipal scale or, at most, at a provincial level. Beginning in 2002, the first aggregations began, culminating then, in the period between 2008 and 2013, with the most recent operations leading to the formation of the main multi-utilities existing today.

In recent years, we have witnessed an aggregation phenomenon that is expected to continue further, stimulated both by the legislative thrust and by management and financial reasons, in the light of the important benefits, from an industrial point of view, deriving from the mere scale involved.

A2A, the company resulting from the merger of the former municipal companies of Milan (AEM and Amsa) and Brescia (ASM), formally started out on 1 January 2008. The desire to combine their strengths and energies and to create value for stakeholders has been expressed in the new name with the tag-line "energy in common" (designed to evoke the idea of projection into the future and a history of unique and exemplary service), other than immediately reflected in actions.

Looking back in decades, the decision taken in 1898 by the Milan City Council to deal with the "electricity issue" and to begin producing energy autonomously led to the birth of AEM (Azienda Elettrica Municipale di Milano—the Milan Municipal Electricity Company). That crucial decision followed the requests, considered exaggerated at the time, made by what was then known as the "Edison Committee". Shortly afterwards, ASM (Azienda dei Servizi Municipalizzati—the Municipal Services Company) also came to life (1908) from a resolution by the Municipality of Brescia, and was entrusted with managing the tram service and the ice factory. With the municipalisation of Milan street-cleaning service and the hiring of around 600 street sweepers who were active at the time, the same period saw the establishment of Amsa (1907). During the following years, Amsa absorbed SPAI (Servizi Pubblici Anonima Italia), which was established in 1929 to provide the waste collection services for Milan. The histories of AEM, Amsa and ASM were interwoven for a century because the communities they served were either overlapped or neighbors, and it was precisely their respective skills that enabled a single, integrated and harmonious business to be created.

2 Relevant Steps After A2A Creation

A2A was the first Italian industrial company listed on the stock exchange market to adopt a dual management and control governance model and tackle the challenge of putting it in place. The Supervisory Board and the Management Board jointly traced out their respective areas of responsibility, setting out exemplary operational rules. This new setting enabled highly specialized companies to be fully integrated on the basis of a clear business plan and common values. As a consequence of the acceleration that was achieved, only a few months later, the Group changed its motto to "energy closer to you" as a means of, on the one hand, emphasizing its closeness to local territories and customer centeredness and, on the other, more effectively communicating its tradition and future aspirations.

The Era of Dualistic Governance (2008–2014)

As mentioned, A2A Spa was established on 1 January 2008 and was the first listed industrial Italian company that, since 2008, has adopted the dual management and control model. Together, the supervisory and management boards have created an operating framework that has allowed full integration, on common values and objectives, of pre-existing companies with strong corporate characteristics.

Post-Merger Integration (2008–2012)

The newborn A2A Group began a phase of internal organizational, operational and corporate reorganization to allow the integration of activities. This process, started immediately after the merger, was very strong in the first 3–4 years, continuing however also in the following years. The year following the merger, a process of external growth also began: in January, the majority of Aspem (multi-utility of Varese) was acquired and, in May, the stake in the Montenegrin EPCG. At the same time, A2A refocused on key businesses and territories, with the sale of its shareholdings in Metroweb (2011), BAS SII (2011), and Coriance (2012).

Edipower (2012–2014)

In 2012, A2A's position in Edipower was consolidated thanks to a series of operations. This path was then concluded in 2015–2016 with the acquisition of 100% of the company's share capital and its subsequent incorporation. In the meantime, the long path of post-merger integration within the A2A Group continued and in 2013 A2A Ambiente was born.

The Relaunch: The 3R3D Three-Year Period (2015–2017)

The new Board of Directors built an innovative Business Plan: the 3R and 3D Plan, based on Restructuring, Relaunch and Redesign, Discipline, Dialogue and Digitisation. The main objective of the Plan was to relaunch and redesign A2A through a strategic repositioning process that would deliver by 2020 a more modern multi-utility, a leader in the environment, smart grid, and new energy models, more balanced and profitable, able to seize the opportunities that will open up in the Green Economy and Smart Cities. Specific M&A operations got under way in sector niches in which the A2A Group wishes to develop its presence: EE (with Consul System) and special waste (Rieco Resmal). Another key phase, which continues today, has focused on territorial aggregations. The main operation of this phase was the partnership with the LGH Group in 2016.

The Transformation: The TEC Plan (2018–Today)

In 2018, the A2A Group approved the first business plan of the renewed corporate governance (in May 2017, after the expiration of the first three-year mandate of the Board of Directors, a new Board was appointed and confirmed Giovanni Valotti as President and Valerio Camerano as CEO). The 2015–2017 three-year plan, annually updated, univocally recognizable in the 3R & 3D strategic plan, has been completed. The new plan, TEC, focuses on three strategic guidelines: Transformation—strengthening and change of reference businesses, broken down into the four business lines of A2A; Excellence—agility of the organization, operational excellence, and process efficiency; Community—attraction and enhancement of the workforce, and full involvement of the external ecosystem. Underlying the three guidelines is Sustainability, the guiding principle behind the development of the A2A Group. In this phase, the process relating to territorial combinations continues with the partnership with the Acsm-Agam Group (2018).

The resulting Group is not yet comparable in size to the main European competitors; even assuming a hypothetical aggregation of the four largest Italian multi-utilities (A2A, Hera, Acea, Iren), the resulting operator would not even be among the top European players in the public services and energy sectors in terms of market capitalization (Fig. 1).

Today, even the less attentive municipal administrations have fully realized that the cost of their own small multi-utility company is out of proportion with the historical benefits, namely greater control and employment leverage: higher operating costs are coupled with a quality of services offered that is often lower than that of companies of significant size. The natural evolution of this vast world of smaller companies is a process of aggregation, which aims to create territorial poles—for simplicity, let us say on a regional basis—where each pole corresponds to a multi-utility that is linked to the territory and can generate added value to be distributed to its stakeholders. The assumptions on which to base the aggregation are the maintenance of the identity and the recognition of existing companies in the territory, the improvement of quality standards and management efficiency, and a boost in infrastructural investments over time and in size.

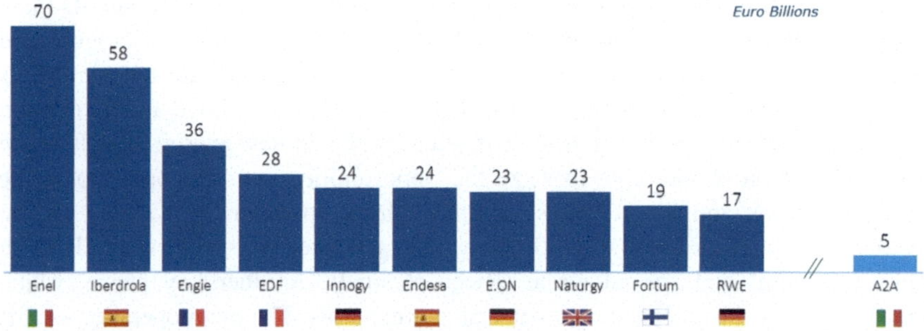

Fig. 1 Stock market capitalisation of the main European utilities of public services and energy as of October 2019

3 A2A Group Business Model

A2A's business model is structured on the basis of areas of specialization. Each Group's company is fundamentally important in achieving the model of efficient city, to which we all aspire. A2A believes that it is only thanks to its deep knowledge of local areas and constant investment in the research and development of new technologies that it can succeed in ensuring the sort of cutting-edge competency that is necessary to guarantee highly efficient, innovative, and top-quality services. A2A's business units are thus the basic constituents of the Group's business model and represent its areas of competency. The uniqueness of A2A is represented by the specific mix of businesses managed. This mix permits to realise threefold benefit from diversification, achieving higher results—see Fig. 2.

Group Mission and Values

The Group aims to provide cities with essential services of the highest quality and efficiency, embodying the principles of sustainability, community enhancement, and positive change. Its strategy focuses on developing a repositioning process that will produce a more modern multi-utility with strong leadership in environment, smart networks, and new energy models, one that is more balanced and profitable, exploiting the opportunities opening up in the Green Economy and Smart Cities. In this process, A2A's values represent its foundation, energy and motivation.

Group Sustainable Innovation

A2A's aim is to develop integrated and cross-cutting projects which look to the future and represent a smart solution for the cities of today, laying at the same time the foundation for the future of cities and territories. The Company pays great attention to research, testing, and technological and sustainable innovation.

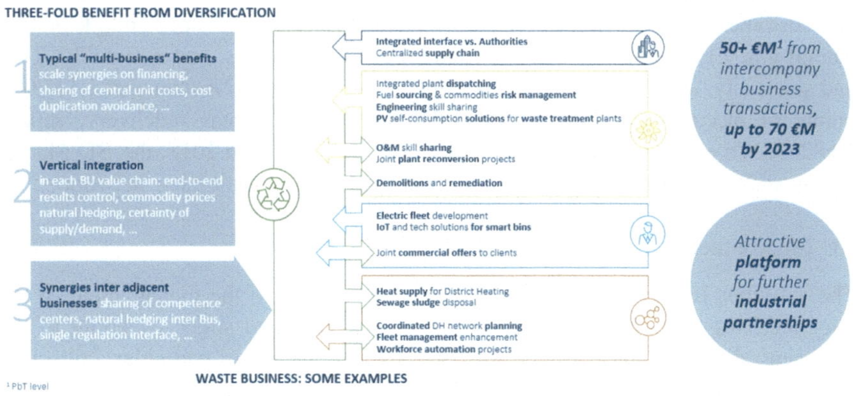

Fig. 2 A2A multi-business value creation

It counts among its projects the construction of innovative plants and systems for the management of energy, activities for the constant improvement of the quality of water, and many more.

Group Business Units

A2A is organized in five Business Units, each following and developing a specific business area.

BU Generation

The activities of the Generation and Trading Business Unit are related to the management of the generation plant portfolio of the Group. The generation sector has the specific goal of maximising plant availability and efficiency, while minimising operating and maintenance (O&M) costs. The "Trading" sector has the task of maximising profit from the management of the energy portfolio through the purchase and sale of electricity, fuel (gaseous and non-gaseous) and environmental certificates on domestic and foreign wholesale markets. The Trading Business Unit also includes the activity of trading on domestic and foreign markets of all energy commodities (gas, electricity, environmental certificates). A2A generation portfolio is made up of ten thermoelectric power plants with 6,694 MW of installed capacity (5,503 MW of CCGT, 886 MW of fuel oil and 305 MW of coal), four large hydroelectric power plants with 1,876 MW of installed capacity (74% of which reservoir) and several photovoltaic plants with about 100 MW installed. Consequently, the total installed capacity is 8,650 MW, with a total electricity production of 17,507 GWh. Moreover, BU Generation manages the trading activities for the entire Group in Italy and European energy markets: power, gas and environmental markets (white certificates, CO_2).

BU Market

The Market Business Unit manages—in direct or in-service mode—sales, marketing and customer service activities related to the supply of commodities and electricity, natural gas, EE, public lighting and traffic regulation systems, and e-mobility, TLC and smart city, addressing all customer segments. In 2018, A2A managed 1,135,000 points of electricity redelivery and 1,511,000 points of gas redelivery, selling 10,826 GWh of electricity and 1,925 Mcm of gas. Finally, in 2018, the BU Market managed 258,090 public lighting points and 2,241 km of optic fiber network.

BU Environment

The activities of the Environment Business Unit relate to the management of the integrated waste cycle, which ranges from collection and street sweeping to the treatment, disposal, and recovery of materials and energy. In particular, collection and street sweeping mainly refer to street cleaning and the collection of waste for transportation to its destination. Waste treatment, on the other hand, is an activity that is carried out in dedicated centers to convert waste in order to make it suitable for the recovery of materials. Lastly, the disposal of urban and special waste in combustion plants or landfills ensures the possible recovery of energy through waste-to-energy (WTE) or biogas production respectively. A2A currently manages nine waste-to-energy plants, 18 biogas/biomass plants and 27 waste treatment plants. In 2018, A2A served 3.53 million inhabitants, collecting 1,671 thousand tons of waste and generating through its WTEs and biogas plants 1,402 GWh_t of heat and 1,807 GWh of electricity.

BU Networks

The main activities of the Networks and Heating Business Unit are the technical and operational management of electricity distribution networks, natural gas transport and distribution networks and the management of the entire integrated water cycle (water collection, aqueduct management, water distribution, sewer network management, purification). It also handles the production and sale of heat conveyed through district heating networks and offers management services for heating plants owned by third parties. In 2018, A2A distributed 11,913 GWh of electricity through 15,571 km of electricity network, 2946 Mcm of gas through 14,218 km of gas network and 72 Mcm of water through 5686 km of water network. A2A also manages a sewage network of 2,567 km. Moreover, in 2018, A2A produced 1,373 GWh_t of heat and sold 2,768 GWh_t, having a total installed thermal capacity of 2,022 MW_t and district heating networks of 1,275 km.

BU International

The International Business Unit manages and develops technological partnerships abroad. Currently, A2A has eight technological partnerships in Europe (Fig. 3).

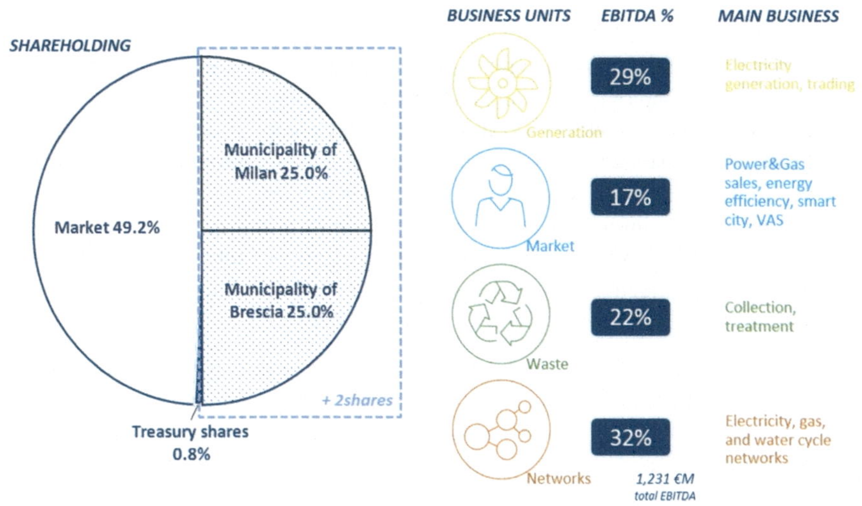

Fig. 3 A2A shareholding structure and business units as of 31 December, 2018

4 New Paths to Territorial Aggregations

In 2014, A2A entered into an intense dialogue with local utilities, starting from the geographical perimeter of Lombardy. The partnership proposal has been summarized in the so-called multi-utility of territories model. This aggregation model has the main characteristic of being based on industrial (and not merely financial) operations, in order to extract both the development and growth potential, and the resulting synergies, and to create a more solid entity to face the current market environment. The model does not foresee a simple merging of companies, but relies on a network or district of companies. This set-up allows the exploitation of economies of scale in businesses with a national dimension, maintaining direct coverage of businesses with greater local importance (e.g. network, collection and sweeping services), ensuring a distribution of investments in the territories and enhancing the identity of existing businesses as well as the skills acquired in the various companies.

The new integrated company resulting from this operation will be operating on a broader territorial scale, generating numerous benefits. For customers, it represents higher quality standards at reasonable prices. For employees, the new industrial development path can provide new career opportunities in a larger group. For territories, new dimensional thresholds that ensure bankability and access to sources of financing provide the possibility of making investments that are not feasible today. For shareholders, higher management efficiency and further development in new territories will have positive effects on profitability and dividends.

Individual companies can ensure proximity to the communities they serve. The companies that share the model thus represent growth poles in the businesses and in the adjacent geographical areas, maintaining leadership in territorial development.

The model is based on seven main pillars:

1. Maintaining the identity and recognition of existing businesses in the territory
2. Improving quality standards and management efficiency levels of services
3. Safeguarding employment levels and local industries
4. Maintaining and enhancing existing brands and points of contact with users
5. Enhancing investment in the territory and recognising autonomy in the criteria for allocating investment in the various business lines
6. Assigning to medium-sized companies the role of a hub for business development and new aggregation projects in the local area of reference
7. Providing subsequent A2A support for growth projects in the "extended" area of reference.

In short, the existing territorial multi-utility companies can grow, improve the level of service to users, invest more and ensure higher dividends to their shareholders. The aggregation of companies, as a whole, improves competitiveness, fosters growth in new territories and, finally, boosts the asset value of the companies.

Multi-utility of territories is a powerful call to collaboration, a key point to strengthen the utilities. The Group considers is a lever to accompany the growth of cities. Cities are human relations, interactions, meetings, being able to connect physically. A smart city cannot be managed by a few stakeholders alone, but must be open to everybody concerned. Secondly, multi-utilities need to share leadership to lessen the carbon footprint; achieving this means nothing less than turning the traditional energy business model on its head! It means that a company like A2A must be as much about saving energy as it is about supplying energy. Decarbonising the power sector is crucial, and A2A is well on the way. Installed capacity in Italy is already over 40% renewable. The next step is to decarbonise other sectors and transport is the main target ($\sim$30% of the CO_2 emitted). Decarbonised electricity paves the way for electric vehicles. Thirdly, transport is a classic urban sector where there has been a great deal of innovation. It is globally driven by city government but for EVs national commitment is necessary; in fact, in all European countries outside Italy, there is quite strong government support for charging infrastructure. This is a developing technology. Development is really rapid but improvements on the ground can be a bit of a step-by-step struggle, so there is a need to help each other in order to make this system work. It is remarkable that interest in this technology is growing by the day; so, to make things better, multi-utilities have to stay together and work together to develop the electric ecosystem.

5 Focus on Distribution Networks and A2A's Contribution

The distribution networks (the widespread infrastructure that reaches households' meters) constitute a natural monopoly. The regulated business translates into low returns which are, however, stable and guaranteed over time. To increase the efficiency in the gas distribution business, the Authority has started a process of concentration of operators through a system of tenders on minimum territorial areas. This new structure will also lead to a further relaunch of investment on local networks. However, currently, gas tenders are struggling to start and, for electricity networks, the deadlines are set for 2030. Even at a technological level, there have not been significant changes: the assets have a ten-year useful life and the transported energy carrier, despite coming from different geographical areas or from different production processes, has not changed: gas or electricity. The impact of global warming on networks is increasing. For A2A, the impact is very concrete because it means a growing stress on urban electricity grids, necessary to guarantee critical services such as hospitals, transport, telecommunications, the integrated water cycle, and also all the economic activities of the territory. It is even more concrete for the operators of A2A's distribution company, Unareti, who, on Thursday 27 June 2019, reported a maximum load on the Milan electricity grid

(1,615 MW), an all-time high for June and not far from the peak of 1,625 MW recorded in July 2015. Another external push comes from macro-trends that are relevant for urban areas. The load that the networks must support due to the electrification of consumption, domotics, IoT, smart home and smart city today has not yet generated such significant impacts as to compromise the performance of the existing infrastructures. Nevertheless, in order to guarantee efficiency in meeting a growing demand, adaptation is essential. There will be no new challenges at a European level for the gas networks either, given the central role of gas in the phase-out of coal, the new international transport infrastructures and the regasification plants under construction. In Italy, the challenge is to reduce dependence on foreign countries, and in particular Russia, to increase the security of the national system. In this specific case, on the gas networks, companies are waiting for the effective emergence of competition.

The urban context, where A2A operates, is even more affected by external dynamics. The most emblematic case is the municipality of Milan. It presents a growing population, new urban settlements, emerging services and an increasing electric commuting from the province (further facilitated by the new integrated fare system in force since July 2019) and from other urban areas of Lombardy such as Brescia (less than 40 min by high-speed train) and Bergamo, which are again attracting residents but, in this case, to the detriment of the provinces.

6 Macro-Development Trends

A transition has begun between utilities oriented towards investment in large-scale plants with the use of conventional energy and business models that will increasingly support a pervasiveness of utilities in urban dynamics. Utilities are the main player of an ecosystem set for a paradigm shift in energy use, one that streamlines connections between market sectors, industries, businesses and community members, with a view towards generating individual well-being and a healthy environment.

Development of Local Demand

We identify the growth of demand as pivotal to the economic strategy through policies aimed at satisfying emerging needs in terms of services, goods and modern infrastructures. This is even more important if it is a multi-utility with control or public participation, as a public finance segment in the broad sense and a subject where the assets and capitals of the community are allocated and which, ultimately, must translate into value generated for the local communities of reference. If this does not happen, the pact with the citizens who have contributed through general taxation to supporting the growth of multi-utilities, is betrayed. A negative example in this sense are some waste management companies with in-house assignments that, while charging high and off-market prices to citizens, often victims of information asymmetry, manage to accumulate losses and do not develop investments for the territory due to the structural dwarfism that distinguishes them.

Improvement of Air Quality Through the Green Economy

While the cities of the 1960s were "imposed", the urban agglomerations of the future will be shaped according to the active demand of citizens, less and less dependent on the policies of local authorities, visible only if they can operate well on a broader scale, like that of the metropolitan areas (Fig. 4).

An example of how local authorities' policies have, even in recent years, caused behavioral changes among citizens, is the effect generated by Ecopass/Area C in Milan. In the metropolitan area, registered cars have decreased significantly in just two years, with obvious benefits both for the city viability and for the quantity of pollutants released into the air. The issue of traffic in urban areas is one of the fundamental elements of urban policies. While, on the one hand, a user decision is needed to abandon the models of traditional petrol or diesel cars, on the other, the public must support a rational territory-wide infrastructure to enable this cultural change, as necessary as ever to curb pollutant emissions in urban areas that are already sufficiently polluted, and to mitigate excess CO_2 production. A city that is less noisy, less polluted and less congested by traffic comes closer to a concept of beauty and harmony, to a people-oriented city. If what really makes a city "beautiful" can be considered a subjective concept, there is an unequivocal element in statistics on health protection: the main cause of death in the coming years, the OECD reminds us, will be particulate pollution of the air (Fig. 5).

Though transport is one of the major sources of pollution, in urban areas heating also represents a major challenge in terms of energy. About half of the air pollution is due to heating boilers: to reduce this percentage, instead of just financing the interventions of individuals that reduce pollution to a minimum (e.g. tax relief for EE work), it would be more useful for local authorities to devote themselves to projects with significant impacts on the community. For example, support for

Fig. 4 Number of private cars circulating in Milan. *Source* ACI data

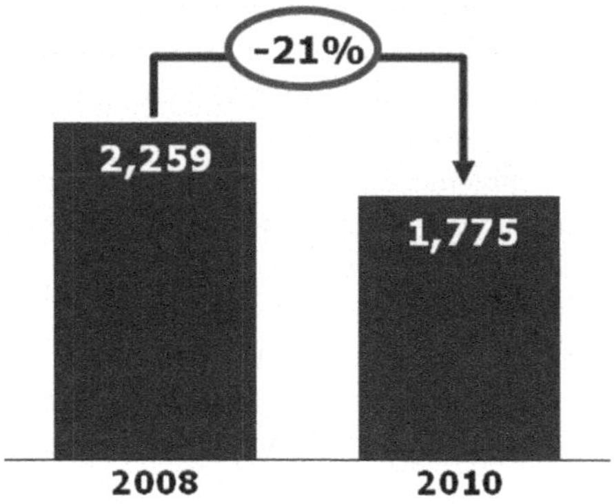

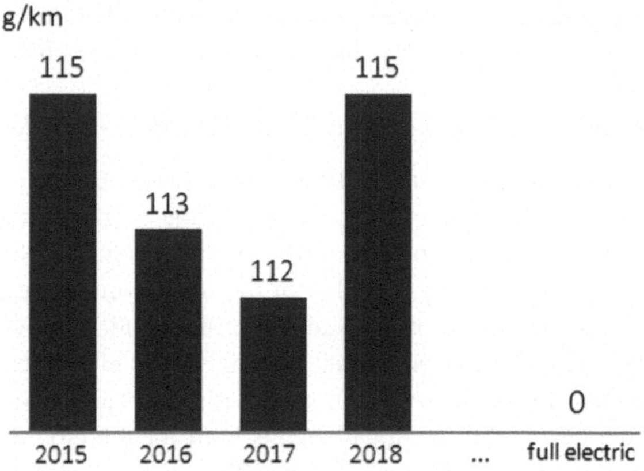

Fig. 5 Circulating cars' average CO_2 emissions in Italy

district heating fueled by a mix of sources recovering heat otherwise lost by production processes, or extract heat from non-recoverable waste in terms of materials, which are then transformed from waste into a real resource.

Innovation: IoT Solutions

The main Italian multi-utilities, if they want to continue to be useful to their territories, must facilitate and push changes in everyday life as, in the last few decades, they have changed the face of Italy, equipping it with infrastructures and services. Traditional infrastructures (e.g. electricity networks, public lighting networks and even optical fibre, now the subject of widespread commercial offers) are the basis on which to embed a new data transmission network. This new network will enable Internet of Things (IoT) solutions, whose development is likely to be slowed down or even blocked by resistance to change both at a public and at a company level.

The technologies, albeit already available for some time, actually see further improvements thanks to a combination of several factors: technological adjustments requested by the regulator (e.g. new electronic gas/electricity meters), development plans for private telecommunications entities, greater cost-effectiveness of services already performed (e.g. heat distributors, water meters), and, the most worthy of attention, emerging services. The growth of a wide range of emerging services, pertaining to the world of smart cities, which, until now, has been more discussed than implemented, can concretely make available to citizens, government and companies a wealth of information in real time, useful for improving the safety and quality of life. This revolution has the germ of innovation, seen as an innovative application of existing technologies.

One of the fundamental assets able to give life to a communication network is the street lighting pole, used as a support for signal transmitters, very similar to radio antennas, and recently the subject of another innovative application. Just as, in the mid-1920s in Milan, the replacement of gas street lamps with electric streetlights was started, today, the process that led to the replacement of lighting with LED technology was completed. The services that can exploit the potential of a telecommunication network are many: field sensors dedicated to such areas as environmental control, measurements of polluting elements, noise, traffic, video surveillance, parking, etc. Added to this are other innovations of which the utilities must become the driving force, such as additional services of EE, thermal, electrical, energy management solutions for homes, demand-response, etc.

Multi-utilities must not look for market space by exploiting the advantages given by the monopoly. The development opportunities, which can respond to the new needs of local communities, come about, first of all, by enabling the intervention of both private operators and mixed-capital entities. By its nature, technological innovation is not imposed by concession, so a territory-wide presence certainly facilitates a central role for operators who manage local public networks and services. Networks are the backbone on which to graft new services, but being open source is a pre-condition for gathering the best skills present in a territory.

For utilities, the market is there; it is a matter of having the courage to pursue the new paths of innovation with determination.

7 Future Scenarios

Short-Term Business Scenario

A2A designed its view for the next five years within three macro-trends: Circular Economy, Energy Transition, and smart solutions. Growth in the waste sector is framed by Circular Economy policies from sorted collection to effective material recovery. The Group's strategy is meant to encourage better waste management in its key territories, to support material recovery, and to structure a reliable market for secondary raw materials. Investment in the Circular Economy from the Waste Business Unit is expected to reach about €670 million in the period 2019–2023. Further actions envisaged include responsibility and sustainability in water usage—with the target of an 18% reduction of water linear losses compared to 2017—and reduction in the percentage of inhabitants non-served by water treatment plants, from the current 22 to 8% in 2023.

A2A's long-term development path in the energy sector is rooted in decarbonisation with the phase-out of coal plants, a key role of gas for energy peak demand and strong growth in renewable resources, aimed at lowering greenhouse-gas emissions, increased EE. This path is supported by market and technological evolution, such as the decrease of LCOE of battery storage, wind and PV. The aim to contribute to national and European targets for emissions reduction through further development of district heating, which will cut NO_x emissions by 260 tons per year and CO_2 emissions by 350,000 tons per year by 2023. A2A aims to achieve promotion of EE in final energy through dedicated offers to retail customers (i.e. high efficiency boilers, LED lamps) and efficiency improvements in the residential, industrial and service industry; it expects a decrease of CO_2 emissions by 6.5 million tons thanks to these initiatives.

As for smart solutions, A2A's view considers compulsory the adoption of digital technologies to make the utilities business more efficient, reliable and safe. In terms of services, it provides municipalities and inhabitants with energy-efficiency projects, IoT solutions, and green mobility. With a decarbonised electricity strategy and growing electrification, A2A paves the way for the e-mobility revolution, recently supported by national incentives and boosted by city government. The aim is to increase the reliability and security of the distribution network with over 140,000 new-generation smart water meters; 3,000 smart parking slots; over 1,660 new charging stations for electric vehicles; 18,000 smart bins for waste collection.

The Role of New Technologies

The new technological solutions can generate valuable services for citizens and for the territories in which local utilities operate, determining new business models capable of supporting the transformation that will take place. A new technology already widely used in the world of utilities is Robotic Process Automation. It is used to replace the need for human work when it is dangerous (for example, using drones to verify the condition of power plants), or in the replacement of activities at risk of human error (such as in repetitive administrative processes involving large numbers of transactions, as in the retail world).

Advanced Analytics are an example of technology that can generate value for multi-utilities: not only by reducing costs within the organization, but also by offering new tools to offer the end customer a customisable service based on their needs. Thanks to the information predictability, they represent a technology that will probably be used extensively by multi-utilities in the future, also due to the possibility of raising the actual production levels and the performance of the operations area. Another technology that can be widely used by companies operating in energy transformation is cloud-computing. This tool helps to integrate the various application systems and convey data from different management systems and business areas.

The digitization of processes has the objective of increasing the productivity, quality, and speed of the information that is transmitted but, also to obtain the advantages of the cloud transformation, it is necessary to perform a set of complex infrastructural activities. The multi-utilities that will be able to re-launch themselves in an agile way in the coming years will also be the ones most likely to embrace the cloud transformation. Research conducted by Gartner on the digitisation of utilities shows that, during the last two years, companies operating in this sector have moved from an embryonic stage to a phase of growth in the use of innovative technologies.[1]

Demographic Dynamics

In the case of local utilities, looking at the territory means bringing out and supporting the positive forces released by other productive enterprises, by associations, by citizens in general, managing to interpret the effects of changes in the demographic structure in advance. Italy is a country where the over 65 make up 23% of a population that is doomed to get older and older.[2] This evolution will lead to growth in the demand for welfare services, whose taxation is the main source of finance today; it is clear that, as the workforce decreases, it will no longer be possible to sustain current spending simply through the fiscal lever. Not even an activity to improve the efficiency of the existing offer will be enough since increasing the tax burden on active subjects is not sustainable: the potentially active population, between 15 and 64, will be smaller and smaller, shrinking from 64% today to 57%[3] in 2050.

Adding complexity to the picture is the population increase forecast in urban areas, where family welfare safety-nets are usually less present. The trend is already in place in Italy: whereas the population in Italy decreased by 0.7% in the last five years, in other large cities it has grown (e.g. in Milan by 4%). The debate on public welfare-state models of financing and disbursement (the so-called Beveridge system) and social insurance with contribution and joint public-private disbursement (the Bismarck system) offers hints towards changing systems at a macroeconomic level, inevitably foreshadowing the entry of the private sector even into the most centralized State systems in order to respond to emerging social needs.

Rethinking the role of multi-utilities in a social sense means adapting to the changes in the existing scenario, without departing from the objectives of company efficiency and profitability. Repositioning is necessary not because of philanthropy or changed managerial sensitivity, but because the business cores that guaranteed the development of multi-utilities (in A2A, about 50% of the gross operating margin is still accounted for by the energy and heat supply chain) become thinner as a result of the change in the structure of supply and demand. Some figures offer clear evidence: in 2008, about 10 years ago, there were 38,000 generation plants in Italy; at the beginning of 2015, the total number of plants in operation with the Conto Energia system (a specific system for photovoltaic and solar thermodynamic) amounted to over 550,000. Many facilities, many of which of extremely small size. It must be remembered that this transformation occurred due to robust incentive policies supported by the resources provided by citizens: the A3 tariff component of the energy bill paid by end customers grew from €3 billion in 2009 to a peak of

[1]Gartner (2019).

[2]Istat, Population at 01.01.2019.

[3]The 2015 Ageing Report—European Commission.

over €14 billion in 2016, before returning to €12 billion in 2018. Forward curves of the PUN (Single Energy Price) provide important signals, constantly updated downwards in the strategic plans of the operators, and by the demand for gas in Italy, constantly falling between 2010 and 2014, with a recovery only in 2015 thanks to the residential sector and, in particular, to the thermoelectric that benefited from the heat peaks of July. Nevertheless, the demand stops at 68.9 Gcm, about 13% less than in 2010.

The economic subjects that, by nature and history, have guaranteed, from the Giolitti law of 1903 onwards, a rooting in the basic services to citizens have been the municipalised companies. From 1903 to the present, they have transformed themselves, and the ambition of each one is, as anticipated in the previous point, to become a multi-utility of the territories, a broad concept that has, among its industrial prerequisites, investments in infrastructures of high technological value for the benefit of the territory. Facing an ageing population, the only possible answer is for the multi-utilities to accompany the change in urban areas, building an inclusive and participatory Smart City model. This is a theoretical premise of the urban communities of the future, currently conceived as careful planning of cities oriented to the needs of citizens in terms of infrastructures, applications and innovations in personal services. The pursuit of social objectives goes by way of an awareness of the changed conditions of the population structure and an ability to respond to new emerging needs.

Hera Group: The Path Towards Shared Value and Circularity

Stefano Venier and Stefano Verde

Abstract Due to the nature of the services they provide, utilities are undergoing a change in several aspects of their strategies when facing new needs from society and new environmental and sustainability challenges. The Hera Group has progressively redrafted some of its reference schemes, so as to tackle new challenges that are arising. Firstly, a move from Corporate Social Responsibility (CSR) to an approach based on Creating Shared Value (CSV) was necessary to respond to the calls from many stakeholders and environmental priorities. Secondly, Hera's services are and will be increasingly focused on circular economy principles in order to further reduce its carbon footprint, water footprint and resource footprint, with the aim of contributing to the fight against climate change and natural resource depletion. Finally, complexity and unpredictability are increasing and need to be tackled with proper enterprise risk management processes, a sound and coherent medium-long term strategic plan and an effective communication with the ecosystem so as to engage with and mobilize all stakeholders.

Keywords Corporate social responsibility · Creating shared value · Circular economy · Enterprise risk management · Strategy · Stakeholder engagement

1 Introduction

Utilities are facing a severe increase in the complexity and speed of change in their businesses, their relationships with communities and territories, and in general in the ecosystem. Over the last 5–10 years, several macrotrends and pressures from the public have emerged and need to be addressed.

This is the case, with the mounting pressure on companies to play their role and contribute to responding to some of the main challenges faced by our planet (i.e. climate change). Pressures come from institutions, civil society and financial markets as well, and utilities need to redraft their strategies to better cope with them.

The Hera Group is facing these challenging times with a comprehensive and well-structured review of some of the key elements defining its strategic process: its Corporate Social Responsibility (CSR) approach is evolving towards a more

A. Gilardoni (ed.), *The Italian Utilities Industry*,
https://doi.org/10.1007/978-3-030-37677-2_8

proactive and comprehensive Creating Shared Value (CSV) paradigm; its business models are increasingly addressing and applying circular economy principles so as to reduce its business footprint; Enterprise Risk Management now ranks high in the company's priorities so as to best manage and plan actions to cope with higher complexity and uncertainty.

One must not forget that the Hera Group provides its services in a vulnerable area of Italy, namely the Po Valley (Pianura Padana), where major economic activities (industry, tourism) and large cities are concentrated; where dust particles such as $PM_{2.5}$ and PM_{10} frequently exceed the maximum thresholds and are addressed with extraordinary mobility policies (bans on car use); where episodes from flooding to drought have continuously risen in recent years. Tackling all these urgencies requires a sound planning and risk assessment process.

Therefore, this chapter will share some of the initiatives we have been developing so far on these issues, since we are aware that these challenges cannot be postponed. Several approaches and solutions are indeed available to utilities and other companies to draft a new path of growth and reap the (economic and social) benefits.

A Brief Introduction to the Hera Group

The Hera Group is Italy's largest multi-utility company by market capitalization, with rapid and uninterrupted growth seen since its establishment in 2002. It has been listed on the Milan Stock Exchange since 2003 (and, since March 2019, included in the FTSE MIB index—the main index of Borsa Italiana) and has the most diversified shareholding among multi-utilities: around 50% of the share capital is controlled by over 100 municipalities and the rest spread over 21k different shareholders.

The Group serves roughly 4.4 million citizens in 349 municipalities covering 5 Italian regions (Emilia-Romagna, Veneto, Friuli-Venezia Giulia, Marche and Tuscany). The main services provided are: distribution and sales of gas and electricity, aqueduct, sewage and water purification services and waste collection and treatment. In 2018, Hera reported €6.1 billion in revenues and had 8777 employees.

The Hera Group's goal is to be the best multi-utility for its customers, workforce and shareholders. It aims to achieve this through further enhancement of its original and sustainable business model, capable of responsible development, innovation and strong interconnection with the areas in which it operates. CSV has become a core focus for mid-term development.

The 5-year business plan to 2022 aims at continuing the Group's development pattern and improving environmental care, in terms of its carbon footprint (23% reduction of the carbon intensity index of energy production by 2022 compared to 2015), energy efficiency (6% reduction of energy consumption by 2022 compared to 2013) responsible use of resources

(73.2% sorted household waste collection by 2022; 76% recycling rate for packaging and 60% overall recycling rate for household waste).

Hera takes part in several international organizations aiming at promoting sustainable development, among which The Global Compact, CDP, the CEO Water mandate and the Ellen MacArthur Foundation CE100.

2 A 10-Year Path from CSR to the New Concept of Creating Shared Value (CSV)

2.1 The Experience Gained with CSR

As far as sustainability and are concerned, the Hera Group was one of the first Italian companies to introduce in its organization a business unit, at CEO level, fully dedicated to these issues. Beginning in 2005, the CSR unit took care of integrating a sustainability perspective within the Group's strategy and business management and reporting yearly on the many responsible actions undertaken by the Group to its main stakeholders.

With the aim of reporting properly to each stakeholder on the efforts made year after year to match its business operations with a sustainable approach, Hera started publishing its Sustainability report as a key means with which to disclose non-financial information simultaneously with financial figures, providing a comprehensive overview for the evaluation of the company's performance. Non-financial information disclosure only became compulsory for European companies with the "Barnier Directive" in 2014 (2014/95/EU Directive), requiring the biggest firms to publish details on their business and environmental aspects, social issues, employment policies and other sustainability-related themes.

The Italian Government converted the Directive into national law only at the end of 2016, making it effective from 2017 onwards. The Hera Group, as one of the earliest movers, was compliant with these requirements almost 10 years in advance [1].

Year after year CSR gained increasing importance within the Group, becoming a key factor, integrated within Hera's planning and control activities. Specific actions made it possible to reach this outcome. For example, with the introduction of CSR targets in the balanced scorecard systems: mechanisms rewarding top and middle management were directly linked to social and environmental performance and sustainable projects to be introduced and fully implemented. Furthermore, a set of key performance indicators and targets referring to sustainability were included in the strategic plan process, so as to orientate business choices from a medium-term perspective, taking the first steps towards a more ex-ante approach on mitigating environmental impact.

Our CSR approach proved to be effective in showing stakeholders the many actions undertaken in terms of environmental or social effects, but it also proved to somehow come short in underlining the link between the purpose of our company, its strategy and social responsibility. Hera's sustainability report was set out in several sections, devoted to each stakeholder: customers, employees, shareholders, investors and suppliers. This structure was very clear and made the report easy to be consulted by the public, but it could also be perceived as a "sustainability show-case" only, overshadowing the existing link between Hera's purpose and CSR and the role of sustainability in designing strategic development.

In this respect, works by Prof. Porter and Prof. Kramer [2] showed how a traditional CSR approach could suffer from its many but fragmentary initiatives: despite the real benefits in social or environmental terms, an excessive number of actions and projects can result in a disconnection from a clear strategic path, and furthermore does not help overcome the trade-off between company and society.

Also considering the changes ongoing in the last 5 years, Hera decided to evolve its approach from a CSR to a CSV concept, taking as a reference framework for the ultimate objectives the United Nations Call to Action (Agenda 2030).

2.2 Rationales Behind a Move from CSR to CSV

As outlined above, CSV is different from CSR in that it embraces a wider approach, taking account of the many overlaps between society and businesses and looking at these overlaps as the main opportunities for win-win solutions to be deployed. Projects and solutions developed under a CSV approach are able to address both society's needs and the economic targets of a given business.

The move from CSR to CSV is necessarily linked to a wider evolution in a company's mindset, namely a move from a reactive approach (CSR) to a proactive one (CSV). Companies can embrace such a change both as a reaction to external calls from stakeholders and as an internal evolution of their mission and their role within society. At this particular time, both external and internal dimensions are very supportive of this change.

Starting with these "external dimensions", several institutions and stakeholders are calling for a new role for companies. First, the United Nations 2030 Agenda clearly set 17 Sustainable Development Goals (SDGs) to be reached by 2030. It furthermore called for the intervention of institutions, civil society and in particular companies, recognizing the need for joint action involving all the players with a stake to reach a scale intervention able to have a significant impact on the biggest challenges.

The European Union is also playing a very active role in promoting CSV. It introduced shared value in its 2011 Communication for a new CSR strategy, fully contributing to fostering the evolution from a reporting approach ("what do we do and how do we do it?") to a strategic and engaging one ("why do we do it?"). Furthermore, in upcoming years von der Leyen's Commission will work towards building up a fully sustainable Europe through a "green deal", and new opportunities for funding will be made available to those companies engaging in adopting sustainable solutions.

Next, civil society itself is asking companies to take the lead in the fight against some of the main challenges to be faced by the planet in the near future (i.e. climate change, depletion of resources, inequality, digital skills to be updated, etc.). The main supranational institutions are not playing the key role they once did, due to the wave of sovereignism and nationalism which is impeding any coordinated action among countries and governments, both at a world-wide and a EU level. Therefore, civil society is looking towards businesses, calling on them to lead this response, as effective actions are urgent and necessary, and cannot be postponed anymore.

Further external encouragement to adopt a shared value approach is coming from financial shareholders. Under the traditional view—in particular in the US corporate framework—financial players are only focused on the short-term financial results of companies and are reluctant to take sustainability issues into account in their evaluations. In recent years, this tenet has weakened and the relationship between finance and sustainability is quickly evolving to overcome the alleged trade-off between economic results and social responsibility.

Finance is now significantly interested in ESG (Environmental, Social and Governance) issues for a number of reasons, as investigated by Eccles and Klimenko [3]. First, several studies have shown a positive relationship between sustainable business strategies and medium-long term returns. For instance, a study by Nordea Equity Research reported how companies ranking higher in ESG management outperformed those ranking at the bottom by up to 40% in 2012–2015.

Secondly, investment companies have such large dimensions that they cannot ignore systemic risks such as those related to climate change. Several publications recognize environmental risks as those having the highest likelihood and the highest impact within a global risk landscape: the World Economic Forum has recently identified 3 environmental risks in the top 5 ranking both under a likelihood rating and under an impact evaluation. Ten years ago in the same ranking there were no environmental risks at all: this is a clear proof of the change in risk priorities during the 2010s [4]. A traditional portfolio differentiation would not hedge against these risks. Hence, investors have started to look closely at ESG and long-term perspectives in their own interest, to protect their investments.

Finally, demand from financial institutions and insurance companies for ESG financial products is quickly rising and brings growth opportunities. The shift in financial operators' view can be found for example in the Blackrock CEO's recent letter to shareholders [5]. In January 2019, Larry Fink focused his writing on purpose and profit, stressing that "purpose is not the sole pursuit of profits but the animating force for achieving them" and that "profits and purpose are inextricably linked". Once again, purpose is seen as the element to drive strategy in the same way as the CSV paradigm is seen as a shift from a reactive to a strategic approach.

When talking about purpose it is straightforward to move from external to internal "dimensions" supporting the CSV standpoint. Companies are called to identify their purpose so as to stick to it, notwithstanding the daily changes and challenges they have to face. More, a clear purpose is able to inspire, engage and attract employees as well. Indeed, workers are an additional stakeholder calling for companies to define a purpose and adopt a shared value approach, to be aware of their company's values, to adhere to them and be sure their work and effort are contributing to some of society's main needs.

Finally, a CSV approach can account for the many links between different companies belonging to the same ecosystem, thus fostering a search for solutions that are profitable to a number of them. As such, the business system itself is also asking for a new way to look at profits and sustainability, so to widen opportunities available to a plurality of companies in a win-win logic [6].

2.3 CSV in Practice: The Hera Group's Experience

What were the main steps in moving towards CSV for a multi-utility like Hera?

Firstly, the theoretical framework was reviewed and we were able to derive our own CSV definition/interpretation, which best fit our businesses, our activities and our social environment. Finding the best definition to be applied is not a mere lexical or theoretical issue, as the correspondence between a CSV description and a company's activities furthers the inclusion of shared value in every fundamental process within the firm (i.e. strategy, rewarding mechanisms, planning and control, reporting, external communication, etc.)

Hence, it is possible to create shared value when business activities generating margins can also contribute to some of the UN SDGs and, therefore, can be effective in promoting sustainable growth. The link with the global agenda's goals allows for a continuous fine-tuning with the UN call for action and thus overlaps may be updated according to global evolutions. Further, the close connection between shared value and SDGs can also be easily translated and applied to define a clear purpose, with advantages both in internal and external engagement. Then, it was necessary to integrate this fresh view in several company processes.

As far as non-financial reporting is concerned, we changed our traditional report structure (stakeholder-based) to a new structure more coherent with CSV representation (Fig. 1). The report is now organized in three sections, devoted to each "driver" we identified as a pillar for activities falling under CSV: "smart use of energy", "efficient use of resources"; "innovation and contribution to development". This new approach allows all the actions and activities Hera carries out under each driver to be reviewed, contributing to 11 SDGs overall (out of 17), which were associated with each driver. Six of these represent the core purpose (7 and 13 for Energy, 6 and 12 for Resources, 9 and 11 for Territory).

As CSV focuses on overlaps between business activities and SDGs, it may be seen as a subset of the wider set of CSR actions as a whole (Fig. 2). Therefore, our new sustainability report also provides a section summarizing all actions and initiatives carried out by Hera that do not fall under a CSV approach but are still relevant from a "reactive CSR" point of view.

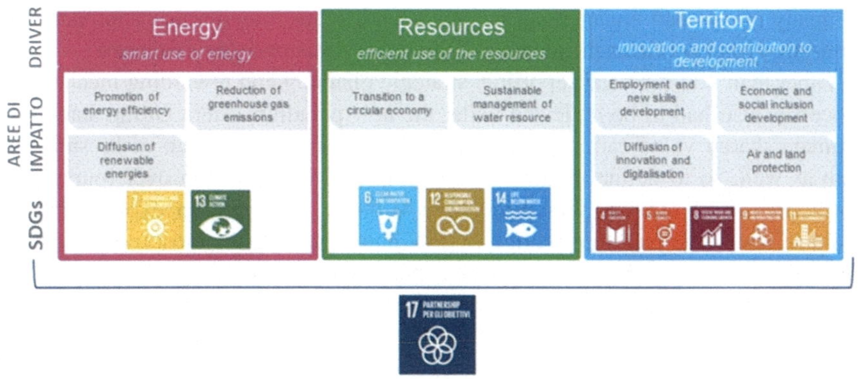

Fig. 1 UN call to action and the reference scheme adopted at Hera

Fig. 2 Existing relationship between CSR and CSV under Hera's approach

A further point to be stressed is the measurement of generated earnings (Ebitda)—and required investments—dealing with activities falling within the CSV perimeter. Since 2016 Hera has adopted a methodology to identify earnings and investments referred to shared value and to account for the evolution of these figures year after year. In 2018 Hera reported a 375 mln € Ebitda coming from CSV actions (36% of consolidated Ebitda), with a significant increase compared to the figures seen in 2016 (300 mln €, 33%). Measuring earnings coming from CSV initiatives brings benefits both in terms of a clear communication and as an easy feedback on company efforts and engagement in implementing the shared value approach [7]. After all, Drucker was right to write "What gets measured gets managed".

Other than non-financial reporting, strategic planning and rewarding mechanism processes also had to be reviewed under the new paradigm. Hera Group started to estimate shared value earnings and investments over the timespan of its strategic plan as well, so to define a path and a target for CSV evolution over four years. Furthermore, all the business units were strongly requested to explore and to evaluate new initiatives contributing to CSV and to include the most robust ones in their business plans. Since the strategic plan is built according to a bottom-up rationale (even though it is inspired by a top-down flow), a sound understanding of the importance and potential of CSV needs to take root within the company so to engage and inspire all key internal stakeholders.

Finally, thanks to the measurement of current and future CSV targets, top and middle management rewarding mechanisms were also updated so as to include shared value objectives in their calculation methods. Nowadays these sustainable targets refer to aspects of shared value, therefore impacting up to 1/3 of the mid-term variable remuneration mechanism for top executives, while 17% of yearly variable remuneration of senior and middle management is linked to shared value targets.

3 Multi-utility Companies and Circular Economy: A Natural and Straightforward Combination

Utilities and multi-utilities provide a variety of services to citizens, most of them essential: water provision and treatment, waste collection and treatment, energy or heat distribution and production are some examples. By the very nature of these services, (multi-)utilities have to deal with the use and regeneration of natural resources as a whole, placing them among companies which can play a key role in promoting and applying circular economy principles [8].

The depletion of natural resources is a challenge that more and more urgently must be addressed: on 29 July 2019 human beings exhausted the biological resources the Planet could renew in 2019 (the so called "Earth Overshoot Day"). As a comparison, in the 1980s the overshoot day was in November and in 2000 it was at end of September. With current production and consumption patterns, the planet is quickly headed towards using its yearly resource budget in just the first half of a year!

Circular economy can be an answer in changing production and consumption patterns and, hopefully, shifting the Earth overshoot day back to the last quarter of the year (or at least fixing it in the third quarter) [9]. The Ellen MacArthur Foundation recently estimated that 55% of GHG emissions refer to energy consumption and are addressed with an increase in RES and energy efficiency, whilst the remaining 45% of emissions comes from the production of good and materials and can be tackled by promoting circular economy in production processes [10, 11].

The first step in Hera's journey towards a circular economy started with the engagement in initiatives and knowledge-sharing within the Ellen MacArthur Foundation, though the CE100 program. Hera has been part of this program, which involves the world-wide best practices across all industries, including academies, since 2017 [12].

Moreover, to put circular economy into practice—either by different businesses and individuals or public administration—we found the 5 Rs scheme very useful as well, that embraces the three core principle of the Ellen MacArthur Foundation.

Under the 5 Rs scheme, the use of natural resources can be decreased by adopting the following behavior/habits:

- "Reduce": consumption can be decreased by identifying those goods or services which do not necessarily have to be bought/consumed and by optimizing the resource's use per unit produced. Our society is used to impulse buying—and this habit has been made easier with e-commerce—and to consuming resources more than is needed (i.e. water, gas, power, etc.). Therefore, the first step involved in applying circular economy principles lies on the demand side, with a significant change in consumption choices and manufacturing processes. Companies and consumers need to reduce their *resource footprint* as well as their *carbon footprint*.
- "Reuse": items are frequently bought and disposed in a very short while. Hence, products should be designed to be reused so as to extend their life-cycle, while consumers need to be made aware of the importance of their behavior and leave a single-use paradigm. This cluster of actions can comprise "Repair" as well, as it is also aimed at extending the life-cycle of goods.
- "Recycle": after having used technical items to the utmost of their lifetime, resources and materials can be recovered with recycling. This is the case for materials which can be reprocessed for producing new items, with savings in the use of raw materials. So far, recycling has been seen mostly as a "mechanical" treatment; in the upcoming years, the need to save as many resources as possible will require us to push ahead with chemical recycling as well (i.e. plastics), with the possible advantage of returning chemical building blocks to be reprocessed (higher use and flexibility).
- "Recover": notwithstanding a reduction in consumption, reuse and recycling, in actual fact a remaining part of waste will not be eligible for being recycled. That residual portion can be used as feedstock to generate power and heating, substituting fossil fuels, with benefits in terms of energy independency and a lower use of landfills as well.
- "Regenerate": finally, after having used natural resources, where applicable it is important to return nutrients and volumes to the environment in the most compatible way (i.e. organic waste in the form of fertilizer or clean water into sources). This final action allows for re-fueling natural sources, contributing both to a reduction in their depletion and to their re-generation.

This versatile scheme can be very useful to any stakeholder to introduce circular economy in their habits, behavior, processes or business models. To provide an example, we will apply the "5 Rs model" to businesses and activities of our multi-utility, to highlight some of the actions we are carrying out to increase our "circularity" and to reduce our footprints (carbon, water and resources one).

3.1 How to Make a Waste Utility More Circular

When applying the 5 Rs model and the three core principles to our waste business, the overlap is almost perfect and, therefore, the contribution to a circular economy can be very significant.

As concerns reducing the production of waste, which is a key target for manufacturers and producers, Hera is providing its citizens with a digital application to sort their waste in the most correct way, resulting in a decrease in unsorted collection. On the same path we are progressively introducing, in accordance with relevant public administrations, new tariff mechanisms rewarding citizens who sort waste better. This change requires an infrastructural roll-out, so as to recognize users when they dispose of their unsorted waste and to measure their disposals (number, volumes or weight). We have developed several tech solutions to provide our territories with best practices in this respect. Furthermore, in recent years the use of Internet of Things and the collection of big data coming from waste bins and dumpsters has also allowed Hera to optimize its collection activities with a reduction in carbon emissions due to a lower consumption of fuel.

Coming to "Reuse", we put effort into communication and training initiatives to increase public awareness on these issues. Furthermore, we have also developed partnerships with non-profit players to allow citizens to donate their used big items (furniture, electric equipment, pharmaceuticals, etc.) with the aim for them to be reused: in 2018, items weighing 800 tons were collected and a 70% was actually reused, whilst the pharma collection allowed for a 500,000 € saving on medical costs. Following the same rationale, we have a partnership to donate fresh meals not consumed from our canteens to non-profit organizations, donating more than 90,000 meals in 10 years.

"Recycling" and "Recovering" actions deal with core businesses in the Waste area. Hera has built, year after year, a large and diversified set of plants to allow for treatment and recovery of urban and special waste, ranging from landfills to waste-to-energy and treatment/recycling plants, following the evolution of society's habits and waste generation. In 2017 we entered the business of recycled plastics production (through the acquisition of Aliplast), allowing Hera to deal with plastics from collection to sorting, mechanical recycling (or energy recovery, if recycling is not possible), and the production of recycled plastic items. The valuable experience we are gaining in the circular management of PET and PE will be applied to more plastics or other materials in the future, increasing our circular footprint.

Another example of the actual application of this concept is the recycling of exhausted vegetable oils collected from households and food shops, to produce bio-diesel used by waste trucks.

Finally, as far as regenerative actions are concerned, the waste business can work on the biological side and contribute by returning nutrients to the soil, converting all the organic waste collected into compost, a natural fertilizer. This process has been enhanced treating upfront the organic waste through a bio-fermentation process so to produce biogas, then used either for [RES] electricity generation or biomethane. The latter represent the most recent development that opened up new opportunities to fuel local transportation and domestic heating inside cities, thus offering not only a green fuel but also a contribution to air quality.

Making waste more circular means making it more resource sustainable and more regenerative.

3.2 How to Make a Water Utility More Circular

Proceeding to the water business, some initiatives we are carrying out are highly representative of a move towards a circular economy.

In the domain of water consumption reduction, Hera is addressing both individuals and companies with specific initiatives. As far as businesses are concerned, our actions include a water management service, namely the provision of solutions to save water in manufacturing or industrial processes, leveraging on the experience gained within the boundaries of the Group where a target of 10% reduction in the *water footprint* has been set. Hera is also targeting a continuous decrease in water leakages in its networks, thanks to several tools: from a "district model" allowing for an optimized management of water pressure to a new digital (algorithmic) monitoring systems, able to highlight network areas more likely to show leakage, to satellite monitoring and detection of network leakages.

As regards households, we will rely on new information to be provided to customers so as to increase their awareness on consumption and saving opportunities. This project will apply to the water business a technique already used in the energy sector, where specific tools based on "nudges" and behavioral economics have been designed. In addition, the water business is an energy-intensive sector and can concretely address an increase in the energy efficiency of its processes to contribute to a reduction in energy consumption.

When referring to water reuse, recycle or recovery, boundaries among these clusters are more blurred than in the waste sector, as water is both a final product for citizens, a raw material and a natural resource. Therefore, all initiatives aimed at extending the use of water and its by-products (i.e. sludge) before reentering the ecosystem should fall within these categories. The water management initiatives provided to our industrial customers are not limited to water savings solutions but include water reuse as a tool for reducing consumption and extending the use of the resource, in a sort of closed loop. Hera has also recently signed an agreement with local public and private stakeholders to use its purified water from Bologna water treatment to support and guarantee the functioning of the two main water canals in the area. As such, when drought or a lack of water occurs, our plant will be able to feed its purified water (up to 600 L per second) to allow a normal and balanced functioning of Bologna's canal network: this is an example of water flow regeneration and surrounding area regeneration as well.

Finally, as for the regeneration of water, all activities dealing with purification of water to be reintroduced in nature are relevant and represent a water utility core business. On this path our Group has recently carried out two large purification projects, namely the Servola purification plant in the Trieste area and a further massive intervention in the Rimini area [13].

Making water more circular means making it more regenerative and more resilient to the effects of climate change.

3.3 How to Make an Energy Utility More Circular

For quite some time, the energy sector has been used to measuring its environmental performance in terms of carbon footprint and carbon emissions, because Emissions Directives and the ETS scheme have been effective in Europe since 2005. The energy sector can address more specifically some of the 5 Rs if we think of energy as a vector, but as any other industry it can approach its processes to include as many interventions as possible.

Reducing energy consumption is the most preeminent target for this industry. The Hera Group is among the utilities who have worked the most on energy efficiency: since 2007, 635 energy saving projects have been carried out both within the Group and at third parties' premises, with a reduction in CO_2 emissions by 1.7 mln tons. The company is targeting a reduction in its own energy consumption by 6% within 2022 (compared to 2013) and several initiatives have been deployed and are producing tangible results: in 2018 versus 2017 the public lighting business reduced its average consumption per lamp by 9%, water purification plants reduced their average consumption per volume of water by 3.8% and a fall by 11.8% was seen in the energy consumption per km of our vehicles.

At third party premises, Hera developed half of its projects eligible for White Certificates since 2007 (the other half being developed internally), totaling nearly 250 projects with a saving near 400 ktep. Two Group companies are devoted to providing energy services to businesses, public administrations or condominiums.

With reference to household consumptions, Hera's commercial offers have been enhanced with the provision of energy saving tools upon request: devices able to allow for remote control and heat and power consumption management; LED lamps able to reduce consumption by up to 80%; the possibility for customers to be provided with renewable gas or power; energy consumption reporting tools. On reporting, the rationale lies on behavioral economics and "nudging", as previously mentioned with reference to the water business: customers are informed on their consumption and that of similar customers and from this comparison they develop a high awareness and attention to energy consumption.

With reference to energy, a distinction between reuse, recycle and recovering is more difficult due to the very nature of this vector. This cluster can include all solutions aimed at maximizing the final use of primary energy consumed (the link with energy efficiency is clear), such as district heating or cogeneration. Hera holds a >500 km district heating grid providing heat to nearly 90,000 apartments and our target is to increase both the grid length and the number of customers served. District heating is a circular solution, as it can exploit hot geothermal water—as in the city of Ferrara—and enhance the use of thermal cascames that would otherwise be lost, thus optimizing energy efficiency and carbon intensity.

As far as regenerative actions are concerned, energy can come from non-regenerable fossil sources and RES (mostly regenerable) ones. The focus of utilities should go to developing RES, so as to cope with circular economy principles. In the years to come, the "regeneration" of green gas from organic waste, from ad hoc agricultural biomasses or from RES produced in excess will rank higher and higher in the EU policy agenda. Making energy more circular means making it more climate compliant.

4 (Multi)Utility Management Features to Face Future Challenges

For a successful development of energy, water and waste businesses, utilities have to look first and foremost at future challenges. We have already referred to resource depletion and climate change as two of the main challenges for utilities, who are called on both to contribute to fight against a further worsening of these macro-trends and to prepare to tackle negative outcomes in case the fight should not succeed or be fast enough. Indeed, utilities are very exposed to the effects of climate change and resource depletion both directly and indirectly.

As utilities generally provide essential services, they need to plan in advance all actions and investments required to guarantee the continuity of their services even under extreme conditions (hurricanes, drought, extreme precipitations, extreme heat, wildfires). They need to be ready to react to unpredictable events both for their social role and for their reputation, and for securing their business targets and future performances. Therefore, risk management and strategic planning play a key role for successful utility management [14].

Enterprise Risk Management (ERM) is a key tool to allow for a comprehensive evaluation of the many dimensions impacted by climate change or resource depletion. Effects coming from these evolutions are complex and inter-related, but ERM is able to map relations, gather additional information and provide top management with sound statistical analyses so as to choose the best investments or strategies to cope with unpredictable events in a medium-long term future. A proper

ERM is also helpful in shifting mitigation actions from an “ex post” approach to an “ex ante” one, where the focus no longer goes to compensating communities or territories for damages occurred, but shifts towards transferring to communities and territories the benefits of a forward-looking management, with the aim of reducing risks, to the advantage of both companies and society.

In the water business, for example, the need for a medium-long term perspective in the development of assets, taking account of possible drought or heavy rainfall situations, is a key success factor. This activity extends beyond a “usual” planning process, as it does not focus on future trends and related likely changes, but it does analyze less likely events so as to allow water infrastructures to be resilient to unlikely events as well. This also goes for Water Safety Plans aimed at guaranteeing appropriate water quality in the long term.

A proper ERM has to be translated into strategic actions within the company business plan. Utilities are used to factoring a medium-long term view into their strategies, due to the very nature of their assets, but they now need to add investments and initiatives able to face not-likely events as well. This is the case, for instance, in Hera’s business plan with reference to grid operations. Within 2022, several actions will be deployed to increase our grid’s resilience to weather conditions: from the reinforcement of the power distribution grid and joint protocols with municipalities in case of power outages, to the installation of smart devices to allow for further remote-controlling and remote-management of our assets in order to react more quickly to disruptions; from the installation of smart meters (some of which with innovative safety functions) at consumers premises to the creation of water grid districts so as to geographically narrow down effects from disruptions.

Finally, the effectiveness of a utility’s strategy in the upcoming years will require a full engagement of their ecosystems as well. Utilities are facing huge challenges and will not be able to tackle them alone, making it necessary to build a strong consensus with other local stakeholders. Communication and engagement will become key activities: communication is necessary to increase public awareness of risks and to encourage correct behavior among communities.

Indeed, utilities can influence citizens’ behavior by providing them with data, analyses and tools to show how current behaviors are badly performing in social/environmental terms and to teach how to modify them with a limited effort. The evolution of some businesses heavily depends on a change in habits: the waste business is the most representative in this respect, as public opinion has to be addressed both to increase sorted collection (quantity and quality) performances, and to become technology friendly so as to fully exploit opportunities from new devices. Furthermore, citizens need to be supported in developing a sound understanding of the waste supply chain, in order to decrease skepticism and a Nimby attitude that is still obstructing infrastructural development in Italy.

To reach such a difficult goal, utilities need to be recognized by authorities, administrations and the population as a reliable business partner, engaged in the promotion of social and environmental quality in the areas served, with long-term targets shared by society. A clear purpose with a social view, transparent non-financial reporting, a set of tangible and fully deployed actions bringing benefits to the territory, a risk management aimed at reducing services disruptions: all of these are definitely tools able to build or reinforce the trust of communities in utilities.

5 Conclusions

Utilities are at the forefront in contributing to the 2030 UN Agenda and can play an important role. To maximize their contribution they need to review some of their processes so to embrace a CSV approach, they must adopt circular economy patterns when modelling their businesses and they are required to apply a risk-management-driven strategy so as to develop a path resilient to changes and challenges.

Over the last decade, the Hera Group has evolved along this path, and this chapter has summarized the main steps undertaken to redraft its processes (strategic planning, non-financial reporting, communication, management rewarding mechanisms) and its business operations towards a CSV and circular economy target point. Our experience suggests the evolution we have undertaken in the 2010s has been rewarding from many perspectives: it has allowed for engaging stakeholders/shareholders, defining our purpose and engaging employees; it has allowed us to match economic results with concrete sustainability goals; it has allowed us to redraw our businesses with a long-term perspective.

It is worthwhile to highlight that technological innovation and digitization are strongly supporting the rise of new circular business models, as they allow for a better citizens engagement, a better sharing of information among players and for continuous advances in R&D to be applied in utility businesses and related manufacturing industries. The climate challenge we are facing is severe, and technologies/digitalization will be a key tool thanks to their exponential development.

In the next decade, utilities will face higher risks and more threats coming from extreme weather conditions and climate change in general. They will need to be ready for this challenge and provide their essential services with no disruption under any event. This is the reason why the Hera Group is undergoing a shift in its risk management and strategic processes, as future success will depend on the investment plans deployed—as of today—to increase services and grid resilience and to develop agile skills necessary to foster such evolution.

The pathways followed by utilities must be fixed well in advance, and our experience suggests that frameworks and tools are now available to set out a path for the 2020s and start thinking of one for the 2030s, with the aim of tracing paths with less footprints.

References

1. Venier S, Bocchi FM (2019) Misurare il valore condiviso: l'evoluzione necessaria nel reporting delle imprese. Harv Bus Rev Ital 3:70–77
2. Porter ME, Kramer MR (2011 Jan/Feb 2–17) Creating shared value. Harv Bus Rev
3. Eccles RG, Klimenko S (2019) La rivoluzione degli investitori. Harv Bus Rev Ital 6:62–71
4. World Economic Forum (2019) The global risks report 2019—14th edn. https://www.weforum.org/reports/the-global-risks-report-2019, last accessed: 2019/09/26
5. Fink L (2019) Purpose and profit—Larry Fink's 2019 letter to CEOs. https://www.blackrock.com/corporate/investor-relations/larry-fink-ceo-letter, last accessed 2019/09/26
6. Youg D, Woods W, Reeves M (2019) Optimize for both social and business value, BCG Henderson Institute. https://www.bcg.com/it-it/publications/2019/optimize-social-business-value.aspx, last accessed: 2019/09/26
7. Porter ME, Hills G, Pfitzer M, Patscheke S, Hawkins E (2012) Measuring shared value—how to unlock value by linking social and business results. https://www.fsg.org/publications/measuring-shared-value#download-area, last accessed 2019/09/26
8. Venier S, Verde S (2018) Una strategia lineare per l'economia circolare. Energia 1:20–25
9. Ellen MacArthur Foundation (2015) Growth within: a circular economy vision for a competitive Europe. www.ellenmacarthurfoundation.org/publications, last accessed 2019/09/26
10. Ellen MacArthur Foundation (2019) Completing the picture: how the circular economy tackles climate change. www.ellenmacarthurfoundation.org/publications, last accessed 2019/09/26
11. Material Economics (2018) The circular economy—a powerful force for climate mitigation
12. https://www.ellenmacarthurfoundation.org/our-work/activities/ce100
13. Venier S (2018) Wastewater management in seaside tourism area: the Rimini seawater protection plan. The Italian Water Industry, Agici, Springer, 225–235
14. Venier S (2019) Cambiamenti climatici ed Enterprise Risk Management. La sfida planetaria, a cura di Enrico Sassoon, Harvard Business Review Italia e Mind Strategiq ed., 97–111

Iren: Growth Through Successful Acquisitions

Massimiliano Bianco

Abstract Iren is an Italian multi-utility company, operating in the sectors of electricity, gas, thermal energy, water and waste management, technology and renewable energy services. Although the company roots date back to the early twentieth century, Iren was founded in 2010, when it inherited the municipal companies of Turin, Genoa, Parma, Reggio Emilia, and Piacenza. These and other acquisitions of smaller companies mainly in the waste and water industries in the subsequent years have contributed significantly not only to the Group's financial growth, but also to making it into a resource for companies in the core territories, due to its technical know-how and financial flexibility. Because of its growth and positive results, especially in recent years, Iren is a case of value creation and distribution for all its stakeholders. In the coming years, Iren Group is expected to grow even further, both organically and externally. On the strength of its consolidated experience and persistent market fragmentation not only will the Group look at small and medium M&A transactions, as it has in recent years, but it will also consider significant deals, leveraging the business model it has developed over the years and its improved financial performance.

Keywords External growth · Diversified portfolio · Multi-utility company · Local communities · Stakeholder value

1 Introducing the Iren Group

Iren is an Italian multi-utility company, operating in the sectors of electricity, gas, thermal energy for district heating, management of integrated water services, waste management services, and technology and renewable energy services. Iren is the main operator in Italy for district heating in terms of volumes connected, third in the water sector for cubic meters managed (2.8 million people served in the water business), and in waste management services in terms of volumes managed (2.3 million people served in the waste cycle).

A. Gilardoni (ed.), *The Italian Utilities Industry*,
https://doi.org/10.1007/978-3-030-37677-2_9

With offices located mainly in Reggio Emilia, Genoa, Parma, Piacenza, Turin, La Spezia and Vercelli, the Group operates with companies organised by business area:

- Iren Energia, generating electricity and heat, managing district heating and technology services;
- Iren Luce Gas e Servizi, sourcing and selling electricity, gas and heat for district heating;
- IRETI, managing and distributing electricity, gas and integrated water services;
- Iren Ambiente, managing the integrated waste cycle, from urban sanitation to separate waste collection, through to designing and managing waste treatment and disposal plants.

Iren is listed on the Italian Stock Exchange (market capitalisation of €3.5 billion) and its majority stake (51.3%) is held by municipalities (mainly Genoa, Turin and Reggio Emilia).

2 From Municipal Companies to Integrated, Cross-Territory Utility

2.1 The Birth of Municipally Owned Utilities

Iren's history goes back to the early 20th century, when the municipally owned utilities of Parma (1906) and Turin (1907) were established. In the following decades, similar companies were established in Genoa (1922), Reggio Emilia (1962), and Piacenza (1972). Despite these companies having several activities in common, such as distribution and sale of gas and electricity, each one had a specific focus, which would significantly influence the consolidation process. In particular, AEM, the municipal public utility of Turin, was mainly dedicated to power production from both thermoelectric and hydroelectric plants, whereas AMGA, in Genoa, was primarily devoted to gas distribution and trade and to the water business. The companies in the Reggio Emilia region, on the other hand, were active also in waste management services (collection and treatment).

For several decades, these companies ran their businesses without significant changes in operations, except for some acquisitions of smaller companies. Only in the late twentieth century did the framework change significantly, particularly in the electricity and gas fields. Following the approval of the European Union, the Italian Legislative Decrees 79/1999 (the "Bersani Decree") and 164/2000 (the "Letta Decree") separated their businesses into essential infrastructure (which constitutes a natural monopoly) and those that can be run competitively. Therefore, businesses running natural monopolies became regulated, while the others were liberalized in order to foster competition among operators. Additionally, elevated public debt, which reduced resources for the central government as well as for municipalities, prompted the latter to reduce their stakes in local public utilities (while remaining majority shareholders) in order to invest in other core institutional activities.

In 1996, AMGA was the first company in Italy to change its legal structure from a municipally owned utility to a corporation and then go public with an IPO on the Borsa Italiana in 1997. In 2000, AEM Turin also was listed on the Borsa Italiana.

2.2 *The AMGA Genova and AEM Turin Merger Creates the Iride Group*

Deregulation and liberalisation led utility companies to deal with the increased level of competition by exploiting all available opportunities of growth, both internal and external. The key strategic moves were:

(a) Industrial and financial consolidation to achieve the size and expertise needed to face the rapidly changing sector;
(b) Development of alliances in order to achieve upstream and downstream integration in the core business' value chain;
(c) Reinforcing the role of vehicle for sustainable development and social responsibility in local territories, considering the technologies in use and the quality of the service offered to customers.

In 2006, AMGA Genoa and AEM Turin merged, creating the Iride Group. The strategic and organisational rationale behind this deal was manifold: to become large enough to develop economies of scale and stimulate further growth; to create an industrial centre rooted in the territory and suitable to stimulate other M&A deals; to optimise the financial structure of the resulting entity, allowing more room to enhance investments; to integrate both the upstream and downstream value chains; and to better balance business risks.

The new entity had 2,841 employees, total sales of €1,400 million, Ebitda of €220 million, and EBIT of about €130 million. It also had a good balance between unregulated and regulated activities. Market activities (sale of gas and electricity) comprised about 10% of Ebitda, while energy production was about 40%. Therefore, regulated business (electricity and gas distribution, integrated water services) amounted to almost 50% of Ebitda. This highly balanced portfolio, a key feature of the Iride Group, was particularly helpful in stabilising cash flows (considering the high volatility in market prices for both production and sale activities) and therefore improving the overall financial risk profile.

The geographic proximity of Genoa and Turin, together with the complementarity of the activities previously run by AMGA and AEM and the focus on core business, strongly influenced the corporate structure. Iride, unlike AMGA and AEM, was not an operating company but a holding company, and its business was run and organised by four companies:

- Iride Energia, based in Turin, managed the production of heat and electricity and the distribution of electricity;
- Iride Mercato, based in Genoa, managed the supply and sale of gas, electricity, and thermal energy. The concentration of every market business in a single entity was aimed at enhancing synergies and optimising back-office activities;
- Iride Acqua Gas, based in Genoa, managed through its subsidiary companies the integrated water and gas distribution business in the district of Genoa;
- Iride Servizi, based in Turin, provided TLC services to local municipalities.

In the first 3 years after the merger, Iride Group experienced solid growth. From 2006 to 2009, Ebitda went from €295 million to €381 million (a 9% compound annual growth rate, or CAGR), return on sales (ROS) from 7.0 to 10.5%, and return on investment (ROI) from 7.1 to 8.2%.

2.3 *The Iren Group as a Result of the Iride and Enia Merger*

As already explained, local utility companies, facing reduction in sales margins and changes in commercial strategies, started a process of consolidation through mergers and agreements. This led to increased bargaining power, economies of scale and, particularly, the integration of supply and production activities (upstream) with commercial ones (downstream). The competitive advantage of aggregation is relevant in the upstream and downstream integration along the energy value chain, where a larger size is of primary importance in the core business areas.

After long negotiations, in 2010 the merger of Iride and Enia took place and the Iren Group was founded. The business and strategic rationale was to create one of the first national operators in the local utility sector, strongly integrated into the gas and energy supply chain, with potential for further organic growth. Moreover, the new group would constitute an aggregation pole, able to stimulate further integration projects, create significant partnerships, optimise the financial structure and reduce financial costs in order to have more room for investments and future acquisitions.

Since its inception, Iren Group has become the leader in Italy in combined heat and power generation and district heating, managing the integrated water business in four districts (Genoa, Reggio Emilia, Parma and Piacenza), distributing 200 million m^3 of water, and managing the waste business, from waste collection to disposal, in three districts: Reggio Emilia, Parma and Piacenza.

In 2010, the Group employed 4,900 people, had total sales of €3.4 billion, Ebitda of €600 million, and EBIT of about €340 million. Due to negative market conditions, in its first years Iren Group experienced only mild growth, with a 3% increase in Ebitda from 2010 to 2013. During this period, the holding company handled strategy, development, coordination and control activities, while its five operating companies ensured the coordination and development of the following business lines:

- Iren Acqua Gas operated in the integrated water cycle;
- Iren Energia operated in the electrical and heat energy production and technology services sector;
- Iren Mercato managed the sale of electricity, gas and district heating;
- Iren Emilia operated in the gas sector, handled waste collection, environmental health and management of local services;
- Iren Ambiente handled the design and management of waste treatment and disposal plants and operated in the renewable energy sector.

Following extraordinary corporate rationalisation operations which became effective 1 January 2016, the activities related to the network's business unit were performed mainly by the company IRETI, which incorporates the former companies Iren Acqua Gas and Iren Emilia. IRETI handles the integrated water cycle, electricity distribution, natural gas distribution and other minor activities. Furthermore, in 2015 the parent company was also involved in a rationalisation project, centralising several functions (such as administration, procurement, legal affairs, personnel, IT). This new organisational structure was aimed at creating synergies through increased economies of scale, centralisation of technical and administrative processes, and the sharing of best practices among businesses and sectors.

3 Iren as a Resource for Companies in the Core Territories

In recent years, a new phase has started for Iren Group. Several deals were closed, mainly in the environmental sectors, i.e. water and waste:

(a) From 2012 to 2014, the Group purchased a controlling stake in Amiat S.p.A., enabling the Group to extend the management of environmental services to the Municipality of Turin in addition to its historical presence in Parma, Reggio Emilia and Piacenza. This meant reaching a total of 123 municipalities and serving approximately 2 million citizens.
(b) In 2015, Iren purchased the Ramo Ligure business unit from Acque Potabili S. p.A., related to the integrated water service in four municipalities in the Genoa ATO, and the water business in several municipalities in the Provinces of Savona and Imperia.
(c) In addition, in 2015 Iren Group acquired control over TRM S.p.A., a company which, among other things, manages the final waste treatment for the province of Turin. TRM has a waste-to-energy plant with a capacity of about 500,000 tons of unsorted urban waste with production of energy. The acquisition enabled the Group to triple its waste-to-energy capacity, confirming Iren as among the top three companies in Italy for waste treatment.

(d) In 2016, Iren acquired control of Atena S.p.A., a utility company leader in north-east Piedmont and active in the regulated businesses of power, gas, water and waste.
(e) In 2017, Iren acquired the Acam Group, which operates in the province of La Spezia, managing the integrated water service (in 24 municipalities with approximately 210,000 residents served), waste management services (in 20 municipalities with approximately 207,000 residents served) and, to a lesser extent, energy services.
(f) In addition, in 2017, Iren acquired GAIA Asti, a company operating in integrated environmental management in Asti and its Province.
(g) In 2018, Iren acquired an equity position in SETA, the concessionaire of the integrated municipal waste collection service in the province of Turin, with approximately 228,000 residents served in more than thirty municipalities.
(h) Finally, in 2019 the Group acquired San Germano, a company operating in waste collection and transport in 145 municipalities for approximately one million residents served in the Piedmont, Sardinia, Lombardy and Emilia Romagna regions, with an annual turnover of approximately €65 million. The company has 20 operating sites and a workforce of more than 800 employees.

These deals, despite being smaller than those described before, have not only contributed significantly to the growth of the Group from a financial standpoint, but have also made it into a resource for the companies located in the core territories, such as Liguria and Piedmont, because of its deeper technical know-how and greater financial flexibility.

In the water sector, Iren now operates in every province in Liguria, and in two of them (Genoa and La Spezia) as the district operator. More importantly, the Group has established itself as a solid vehicle for improving the quality and sustainability of utility services, as demonstrated with Acam, the La Spezia public utility company. Acam was in serious financial distress (it went through a debt restructuring in 2013), which negatively affected the quality of its services and its ability to comply with the high-quality standards demanded by customers. In addition, the province hosts many highly visible tourist attractions and, according to the local regulator, the investments needed in integrated water facilities were about €7 million per year (about €30/person). However, water losses were high, at 56%.

The aggregation allowed Acam to face its financial difficulties (Iren paid Acam's bank debts). It also gave Acam additional financial resources, allowing it to meet important investment goals. Annual investments went from €7 to €20 million, which amounts to €100 per resident, in line with the standards of other European countries. Moreover, Acam benefited from Iren Group's expertise. Iren's best technologies, developed through decades of experience, were made available to help improve the quality and environmental sustainability of Acam's services. Water network projects, an approach widely used by Iren Group (especially in Reggio Emilia) to reduce water losses, were also adopted in La Spezia, with a technique that involves dividing the networks into small, equal areas, or districts,

	2015	2016	2017	2018	CAGR 15-18
Revenues, €mil	3,094	3,283	3,697	4,041	9%
EBITDA, €mil	678	814	820	967	13%
EBIT, €mil	347	427	420	530	15%
Net result, €mil	140	185	265	273	25%
Net financial position / EBITDA	3.20	3.02	2.89	2.54	
Investments	268	270	357	447	19%

Fig. 1 Iren Group's key financial results from 2015 to 2018

which allows daily monitoring and constant analysis of hydraulic parameters. This way, the fundamental campaigns that search for leaks are more accurate and targeted only at the districts where monitoring has revealed hidden leaks.

Significant improvements were introduced in procurement (due to economies of scale), in the processes reorganisation of processes, as well as the sharing of front- and back-office personnel, thanks to the geographic proximity of Genoa and La Spezia. The expertise shared with the waste facilities and the newfound financial stability allowed the Group to increase its portfolio of waste management services (both collection and treatment) in a core territory like Turin. From 2014 to 2018, separate collection in Turin increased from 42 to 46%, and door-to-door collection was partially adopted.

In a recent paper by McKinsey and Company,[1] the authors argue that pursuing "many small deals that accrue to a meaningful amount of market capitalisation over multiple years instead of relying on episodic, 'big-bang' transactions" creates value for large corporations. The paper also states that:

> programmatic acquirers have built up organisational infrastructures and established best practices across all stages of the M&A process—from strategy and sourcing to due diligence and integration planning to establishing the operating model. [...] The data confirm that programmatic acquirers continue to perform better than industry peers; indeed, the more deals a company did, the higher the probability that it would earn excess returns. [...] Precisely because these companies are doing deals systematically, we believe they are building lasting, distinctive capabilities in M&A.

It seems Iren's case perfectly confirms these findings. Figure 1 shows Iren Group's key financial results from 2015 to 2018.

Ebitda grew from €678 million to €967 million, with an CAGR of 13%. Operating profit grew at an 15% CAGR. The net result almost doubled, going from €140 million to €273 million. Investments grew significantly, from €268 million to €447 million, of which €165 million refer to the integrated water service.

[1] "How lots of small M&A deals add up to big value" by Jeff Rudnicki, Kate Siegel, and Andy West; July 2019.

4 2019–2024: Iren Group Scales up to National

The growth strategy undertaken by Iren for 2019–2024 is influenced by the main trends expected through 2030, identified in the business plan issued in 2018 and confirmed in the 2019. These trends are energy transition, centrality of customers, sustainable development, and technological revolution. In particular:

- Customers: the path undertaken in past years is confirmed with increased vigor, focusing on customers/citizens through product offerings and new technology services (such as e-mobility and digital payments).
- Organic Growth: €200 million, mainly generated by integrated water services, development of waste treatment structures, growth of the customer base, district heating and participation in gas tenders. Cumulated capex is €3.3 billion and two thirds of the investments planned, or €2 billion, are aimed at regulated sectors to strengthen, modernize, and digitise network services, with a particular focus on water treatment plants in the Liguria area, electronic gas, water and electricity meters, network resilience, loss reduction, and increasing technical quality.
- Efficiency: the positive trend seen in recent years (€200 million of synergies since 2010) will continue.
- Sustainability: Iren's commitment to protect the environment and fight climate change is made concrete through the circular economy, minimising resource consumption, decarbonisation and the creation of increasingly resilient cities, through the services it offers and products with lower environmental impact.
- People: managing people goes hand in hand with the Group's transformation, with a heavy focus on developing skills through training and retraining programs, professional growth and incentive tools. A continuation of the generational turnover is expected, with the addition of new qualified people entering the field in higher numbers than those expected to leave.

This growth should be achieved organically as well as by seizing external opportunities. As in recent years, on the strength of its consolidated experience and due to persistent market fragmentation the Group will seek small to medium M&A transactions (which could contribute up to €100 million in Ebitda). But it will evaluate larger transactions as well, using a business model developed over many years and thanks to improved financial ratios.

5 Creating and Distributing Value for All the Stakeholders

For more than a century Iren has been providing its territories with public utility services, using its skills, technologies and resources to create value, with the aim of improving the quality of life for people, making companies more competitive and contributing to the growth of the regions in which it operates. Value creation occurs through different perspectives: economic, environmental, and social and cultural.

5.1 Economic Value Creation

Iren's economic and financial performance shows, via a business portfolio with mainly regulated activities and a significant increase in investments, a strong trend towards the development of infrastructure to serve the growing economic systems of the areas served. This confirms Iren's focus on environmental sustainability, digital transformation and innovation, in synergy with the businesses and institutions of the reference areas.

Iren Group produces value using production factors effectively, with the objective of generating added value compared to the external resources used. Furthermore, the business conducted contributes to the economic growth of the social and environmental context in which the Group operates, and produces significant indirect effects on local areas, especially in terms of employment and investment.

Other than economic effects, the activities of the Group also generate important positive environmental impacts through local development, development of basic infrastructures (electricity, gas and water system networks, water treatment plants and sewage systems) and essential services (waste collection and disposal). Moreover, all the business areas present significant opportunities for the development of innovative technologies and processes and, consequently, for territorial growth, also in terms of know-how. The Group contributes to employment in the areas where it operates and generates added value through actions aimed at increasing professional skills and consolidating improvements in education. The Group has no specific local recruitment policy. However, given the specific features of the Italian labor market, nearly all new recruits reside in the province where their place of work is located, while almost 82% of senior managers reside in the same region as their place of work.

This indicator that highlights the Group's ability to produce value within the area and, at the same time, satisfy the economic interests of its main stakeholders is represented by added value. This parameter measures both the economic performance of management and the ability of the Group to generate the conditions necessary to distribute wealth to stakeholders. In 2018, Iren Group generated a total gross added value of €1363 million, up by 6.7% compared to 2017.

The added value created by Iren Group was allocated as follows:

- 35.8% to the Company (approximately €488 million). This is the share of wealth kept within the Group, including depreciation and undistributed profits;
- 29.7% to Personnel (around €405 million). This is the portion made up of salaries and wages, expenses and other personnel costs;
- 12.6% to Local Authorities (around €171 million). This is the portion distributed in the form of direct and indirect taxes, net of the grants received for the year;
- 10.9% to Financial Backers (around €149 million). This portion includes all the financial charges the Group owes to creditors;

€million	**2018**
Revenues from goods and services	3,698
Change in work in progress, semi and finished products	10
Other income	191
Production revenue	**3,899**
Raw materials, consumables, supplies and goods	(1,387)
Cost for services	(1,229)
Other expenses	(20)
Capitalized expenses for internal work	33
Provisions for risks	(81)
Intermediate production costs	**(2,684)**
Gross added value from core business	**1,214**
Non core and non recurring items	148
Total gross added value	**1,362**

Fig. 2 Added value created by Iren Group in 2018

Fig. 3 Iren share price from 2014 until the end of 2018

- 10.3% to Shareholders (over €140 million). This is the portion allocated to shareholders in the form of dividends;
- 0.7% to the Community (around €10 million). This is the portion allocated to local communities through the participation in the development of social, environmental, cultural and sporting events (Fig. 2).

Shareholders also received a significant return thanks to the appreciation of Iren shares. From 2014 to 2018, shareholders realized high returns: Iren shares increased by almost 130%, with €213 million in dividends distributed (see Fig. 3).

5.2 Environment

Environmental protection, rational use of natural resources, and sustainable development have always been central to Iren Group due to the nature of its business and the focus of its mission. This commitment is made clear in the Integrated System Policy, which is distributed and shared with all Group personnel and companies. Respect for and protection of the environment, rational use of water resources, efficiency and reduction in energy consumption, development of RES and proper management of the integrated waste cycle are fundamental elements that direct the Group's strategic choices.

Iren Group undertakes, with responsibility and with the use of human and financial resources, to reduce its impact and protect the environment. The expenses and investments incurred in 2018 for environmental protection amount to over €411 million, of which: 61% to increase the efficiency of electricity and gas distribution networks and water treatment and purification plants, obtain hydroelectric green certificates, and carry out other improvement projects (e.g. smart cities); 32% to optimise separated waste collection systems to pursue the waste recovery objectives set out in the area plans, specifically the extension of the door-to-door service in the Municipality of Turin; 6% to raise the efficiency of the electricity and thermal energy production plants via heat storage, work to make them more flexible and revamping and developing production from RES; 1% to implement services and products with positive environmental impacts for customers (e.g. the e-mobility project).

The Group carefully monitors atmospheric emissions (measurements on chimneys, indirect calculations, number of leaks etc.) in order to identify specific measures to reduce them and verify the results achieved on a regular basis. The generation of electricity from RES creates significant positive effects on the reduction of emissions, and the predominant cogeneration framework (production of electricity and thermal energy that feeds the district heating networks in different cities) of the Group's thermoelectric plants significantly contributes to limiting specific greenhouse gas emissions.

In order to reduce pollution, only natural gas is used to supply the energy production plants and both low-emission combustion systems and pollutant reduction systems are installed. Continuous emission monitoring systems make it possible to detect in real time the main pollutants and the improvement of the efficiency of the combustion process of combined heat and power generation plants, larger thermal plants and waste-to-energy plants.

The percentages of separated waste collection achieved in 2018 show a positive trend across all areas, having reached 70% in the provinces of Parma (from 78.4% in 2017 to 79.2% in 2018), Reggio Emilia (from 69.8% in 2017 to 74.8% in 2018) and La Spezia (70% in 2018). The average for separated waste collection across the areas served by the Group was up compared to the previous year, reaching 64.3%, against the national average of 55.5%, a figure that is very close to the target of 65% set for 2035 by the European Union's Circular Economy Package.

The Group's power production fleet mainly consists of hydroelectric and photovoltaic plants that use RES and cogeneration thermoelectric units generating power via combined cycles, one of the most efficient technologies available on the market. Furthermore, combined heat and power generation plants connected to the urban district heating network provide reduced energy consumption and improve environmental performance compared to traditional heating systems. In total, almost 87% of energy production (compared to the national average of 35%) is powered by RES (hydro or solar) or integrated (cogeneration).

The Group has undertaken many initiatives to reduce the environmental impacts of its main activities:

- Environmental services: all Group waste-to-energy plants are equipped with emission monitoring systems that check both emissions from the energy and heat production plant and the waste combustion process. The monitoring system is guaranteed by continuous measurements and checks on compliance with legislation and the Integrated Environmental Authorisation with the control of the indicated substances.
- Integrated water service: the initiatives aimed at reducing the environmental impacts mainly concern (a) the reduction of energy consumption by adapting wastewater treatment processes and replacing old machines with the latest generation equipment that consumes less energy; (b) the replacement of submerged electric pumps in the pumping stations with new pumps fitted with inverters; (c) reduced water procurement through the reduction of water main leaks; (d) the improvement of the quality of the water that leaves the treatment plants and the connection of stretches of untreated sewage to final treatment systems; (e) the abatement and containment of odorous emissions from treatment plants by confining them to secure rooms during the treatment process in order to allow the air to be aspirated and treated. To reduce leaks in the water networks, the Group has launched a "division into districts" project. Currently, 46% of the total network is divided into districts and, by 2023, it is expected to reach a coverage of about 85% of the network served. The division into districts also produces benefits in terms of energy consumption reduction. In 2018, it allowed savings of approximately 400 TOE. The public water dispensers for the free distribution of drinking water (chilled or sparkling) to the residents made it possible to considerably reduce the use of plastic bottles (approximately 20.6 million 1.5-l bottles in 2018) and, therefore, the production of waste (722 tons of PET avoided). Savings of 1,878 tons of CO_2 can be estimated for 2018, thanks to the non-consumption of 1,371 tons of oil equivalent for bottle production.

5.3 Social and Cultural Capital

As of 31 December 2018, 7,042 employees worked for Iren Group, with a 56% increase compared to 2014, thanks to appointments and aggregations. Employees with a permanent contract or, in the case of young people, an apprenticeship contract represent 99% of all staff. The 1,807 women employed by the Group represent approximately 26% of total employees, a figure higher than the Italian comparable 18% average.

Iren Group has also been devoting significant resources to training, a crucially important tool in growing and enhancing human resources. Alongside the development strategies and values of the Group, training plays a fundamental role, both in pursuing objectives and in meeting the needs of innovation and change that the market dictates to be competitive. Training is focused on maintaining and developing skills, with particular reference to the "core" skills of various professionals who work within the organisation, whether they are technical, specialist, or managerial skills, with a view to lifelong learning.

From a cultural standpoint, Iren Group focuses heavily on cultural, environmental and sports events that possess distinctive characteristics at a national level and which are rooted in the local tradition. The events generally involve cultural and innovation projects, including support for theatres, scientific events, and cultural events; entertainment projects in local areas; projects in the environmental sector, including support for environmental projects and the arrangement of playgrounds and small green areas; projects in the sports sector; and projects in the social sector.

For many years now, Iren Group has been committed to education, the most effective and strategic means of asserting a culture of sustainability and innovation. The Edu. Iren project represents a starting point, which makes available a catalogue of free training proposals for students and teachers on sustainability, water, energy and environmental topics. For instance, in 2018, Edu. Iren involved 78,893 people and over 560 schools in its training and educational projects. The range of educational courses has been extended beyond the traditional school year to make it available to the many summer programmes for younger people, and adult training courses have been intensified.

In 2018, Iren offered more opportunities to experience and get to know the plants. The Parco Acque Depurate (PAD) of Mancasale in Reggio Emilia is the main plant serving the Reggio Emilia area and the first example in Emilia Romagna of the reuse of treated water in quality crops. It has been transformed into an eco-industrial park, open to the public, with an educational mission, where work and cultural activities can exist side-by-side. The park has a 10-stop pedestrian path with informative signs and highly-accessible maps that facilitate the guided tours provided by the employees who run the plant. The Turin waste-to-energy plant continues to be one of the few facilities to offer an innovative educational course and was visited by almost 5,500 people in 2018.

6 Conclusions

Since 2010, Iren Group has changed significantly. It has reshaped its organisation to improve synergies and enhance productivity and quality of services and, through several small and medium acquisition deals, it has increased its business portfolio, strengthening its position in the core territories.

In particular, since 2015, Iren Group has been delivering strong results in terms of value created for all its stakeholders:

(a) The financial results showed a double-digit CAGR of 18% in net results, with Ebitda up by almost €300 million in 4 years (from €678 million in 2015 to €967 million in 2018); shareholders realized a 128% share price increase from 2014 to 2018, also receiving €213 million in dividends;
(b) Environmental sustainability has improved, thanks to the technologies adopted, the effectiveness of actions and practices in use and high investment carried out (especially in the waste and water service sector);
(c) Human resources are a fundamental resource for growth, and that is why around 99% of personnel are hired with permanent contracts or apprenticeship contracts. Iren Group has been devoting significant resources to training, a crucially important tool in growing and enhancing human resources;
(d) Significant value has also been created for communities, thanks to cultural and innovation projects, entertainment projects in the local areas, projects in the environmental sector, and projects in the sports and social sector.

In the coming years Iren Group is expected to grow even further, both organically (thanks to high capex and synergies) and externally. Not only will the Group look, as in recent years, at small and medium M&A transactions supported by market fragmentation and consolidation experience, but now it will also consider major deals, thanks to its business model developed over the years and improved financial ratios.

Acknowledgements The Author wishes to thank Francesco Meringolo for his support with this project.

CVA: Renewable Sources Value Chain in the Experience of a Leading Hydroelectric Player in Italy

Enrico De Girolamo

Abstract This chapter describes the changes undertaken by a leading hydroelectric company in Italy since its foundation in the year 2001. The common thread of regulatory developments will lead us through the diachronic journey that has characterized the European and national scenario from a nationalized sector to liberalised electric markets. The Company timeline has its origin back to the nineteenth and twentieth centuries with the emergence of the steel industry, which required a large amount of energy, hence the development of hydroelectric power plants in the Aosta Valley. It is nearly a century later when the regional government acquired the entire share capital of a company owning three hydroelectric power plants in the region, changing its name in "Compagnia Valdostana delle Acque". By the end of the twentieth century the incumbent state-owned operator sold to the regional government 26 hydroelectric plants located in the Region. Over 20 years CVA has extended to wind and solar sources, becoming the 6th Italian producer of total RES sources and the 4th of hydroelectric source. In 2019 the Company has published its first sustainability report, embracing decarbonisation strategies and innovation with a number of initiatives.

Keywords Renewable energy · Sustainability · Energy Transition · Energy community

1 The Origins of CVA with the Liberalisation of the Electric Market

The Valle d'Aosta region, situated in the north-west of Italy, is generously endowed with a large amount of natural resources: copious water in rivers and glaciers, wide forests, and ore-rich deposits. This combination of factors favoured in the nineteenth and twentieth centuries the establishment of industrial plants, during an historical period characterized by energy-intensive production and limitations in energy transmission systems. Thus, the Valle d'Aosta region turned out to be the ideal site for steel mills interested in exploiting local mines. Requiring a large

A. Gilardoni (ed.), *The Italian Utilities Industry*,
https://doi.org/10.1007/978-3-030-37677-2_10

Fig. 1 Dam's Construction—Beauregard. *Source* CVA

amount of energy, these companies developed the original core of hydroelectric power plants in the region (Fig. 1).

Even after the end of local minerals exploitation, following the increasing development of the industry in Italy, the Valle d'Aosta region continued to play an important role for energy production. In fact, many hydroelectric power plants were constructed at the beginning of the twentieth century, as well as important dams, still perfectly operating.

The first crucial step in CVA's history was the acquisition by Valle d'Aosta's regional government in 1995 of the entire share capital of a local company owning and operating three hydroelectric power plants. Over time, the company—whose name was changed to "Compagnia Valdostana delle Acque"—grew by adding further power plants through several transactions. In particular, the process of privatisation and liberalisation of the Italian energy sector gave the company the opportunity to acquire other hydroelectric assets located in the region. In 2001, the then state-owned Italian operator, Enel S.p.A., transferred to one of its subsidiaries, Geval S.p.A., twenty-six hydro power plants, of which twenty-five were located in the Valle d'Aosta region and one in Piedmont, close to the regional border. Thereafter, through its holding company Finaosta S.p.A., the regional government bought out Geval and merged it with Compagnia Valdostana delle Acque S.p.A. to form "C.V.A. S.p.A.—Compagnia Valdostana delle Acque—Compagnie Valdôtaine des Eaux S.p.A.".

Between 2001 and 2009, CVA primarily focused on the management of its existing fleet of power plants, while gradually consolidating its position in the energy market through the activities of its energy trading and management subsidiary CVA Trading, established in 2001. Over time, CVA transferred its sales contracts to CVA Trading, in pursuit of a clear, functional allocation of business activities within the Group: energy production, sales, and trading activities.

The deal of 2001 also included the acquisition of half of the share capital of Deval S.p.A., which sold and distributed electricity to end users. In a series of transactions throughout the following decade, CVA completed the acquisition from Finaosta and Enel of their respective shareholdings in:

- Deval S.p.A., which meanwhile became the prevalent operator of the concession for power distribution grid in the Valle d'Aosta region, and
- Vallenergie S.p.A., which focused on electricity sales in the Valle d'Aosta region to retail customers in the Regulated Market and which, in 2013, was merged into the existing CVA Trading.

Since 1 January 2014, CVA Trading has become the sole utility in the Valle d'Aosta region with authorization to serve the Regulated Market. In 2016, this service has been provided by the Group under Enerbaltea trade name and brand accomplishing regulatory requirement on unbundling. In light of the Italian government's decision to phase out Regulated Market power sales, which are scheduled to end altogether in July 2020, the Group has accomplished regulatory requirement on unbundling by clearly distinguishing between the services it provides to the Regulated Market (under the Enerbaltea trade name) and the services it provides in the open market (through the CVA Trading trade name).

The separation between distribution and selling comes from the unbundling rules, introduced by the 96/92/EC Directive of the European Parliament and of the Council dated 19 December 1996, concerning common rules for the internal market in electricity. As a result of this Directive and, particularly, of its transposition into Italian laws, companies or groups of vertically integrated companies that operate simultaneously in the infrastructure sectors of electricity and gas (distribution) and in free sectors (production and sales) are required to fulfil the functional separation obligations. In fact, the idea is that if a single company operates a transmission network and generates or sells energy at the same time, it may have an incentive to obstruct competitors' access to the infrastructures. This would prevent fair competition and presumably increase prices for end consumers.

Hence, the liberalisation of the Italian market has led to the general transformation of this sector, from a single national centralised system, managed under a monopoly regime, to the creation of multiple actors, competing with each other on a levelled playing field. CVA has benefitted from the liberalisation opportunity: nowadays its operations span along the entire value chain from power generation to retail sales, with dedicated internal distinct companies.

2 Corporate and Business Structure: Major Steps and Future Prospects

The energy produced by CVA comes entirely from RES: water, wind and sun. These three elements give their contribution to a complete clean energy production. Hydro source represents over 10% of the Italian energy production and CVA contributes 6.4% in terms of volume of energy generated, counting as the fourth largest hydroelectric energy producer in Italy.

History left in CVA hands the privilege, as well as the duty, to manage a heritage of over 1 GW of hydroelectric power plants, which produces every year around 3 TWh from 32 power plants of different size, each requiring high levels of maintenance to guarantee the highest standards of performance for the company and of safety for workers and the territory. For instance, hydroelectric generation, especially when combined with large reservoirs, has a key role in flood control: some power plants are equipped with the capability at the watercourse intake to modify the flow rates used by the power plant to a different level than natural discharge.

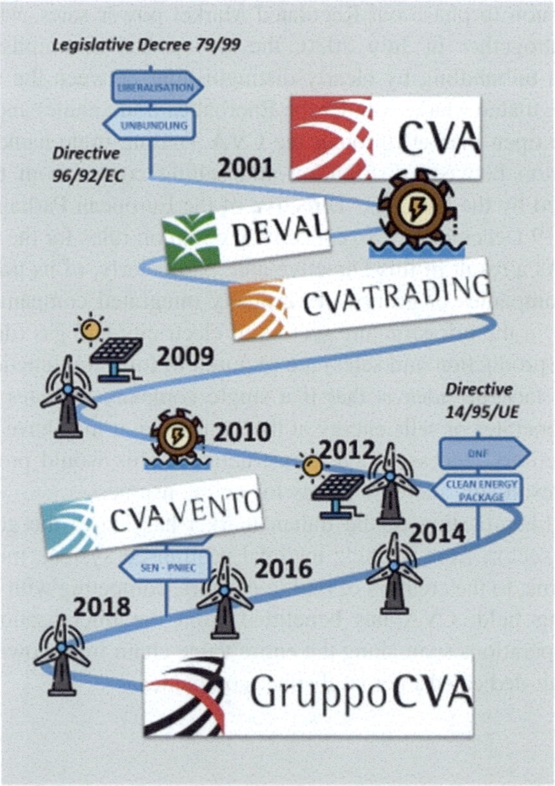

CVA owns six large dams that are supervised by the Directorate General for Dams and Water and Electrical Infrastructure of the Italian Ministry for Infrastructure and Transport (Figs. 2 and 3). CVA is required to carry out the mandatory control, inspection, and maintenance activities on the dams, as well as on the banks and reservoirs, in order to confirm the proper working condition and maintenance of the structures and ensure the safety of the population residing downstream. Therefore, CVA is fully aware of the importance of a correct management of its hydro assets with regard both to people who live in the areas interested by the plants and to the surrounding environment. In addition, during recent years CVA has extended its interest for clean and RES also to solar power plants and wind farms.

The road to full renewable production has been strictly followed by CVA and its subsidiaries and led the Group to gradually expand, till 158 MW of wind farms and 12.5 MW of solar owned today.

The history of CVA mirrors that of Valle d'Aosta, where the Company Hydroelectric capacity is generated. Among the Italian Regions, Valle d'Aosta is an ideal natural territory for the development of sustainable practices: from the deployment of renewable energy to the promotion of the regional environmental assets. The regional government is also the legal owner of the Company by means of its financial body Finaosta. To this extent, the Company's industrial strategy towards sustainable development is naturally tuned to that of the regional government.

The actual European and national scenario is such, that once again CVA finds itself in the position to hand over and develop a significant leadership in the field of

Fig. 2 Goillet Dam. *Source* CVA

Fig. 3 Place Moulin Dam's Basin. *Source* CVA

renewable sources from the small region of Valle d'Aosta towards a national and international playing field.

Since the establishment of CVA in 2001, European Climate targets have progressively enhanced the central role of renewable sources in the European energy system, setting up a new pace for entrepreneurship in the renewable field. A number of regulations over the past two decades have punctuated a precise roadmap. Of the three thematic priorities established by the 7th Environment Action Programme to lead EU environmental policy until the end of 2020 [1], that of turning the Union into a resource efficient, green and competitive low carbon economy and the goal to secure investment for environment and climate policy, have foreseen a leading role for the renewable energy sector, therefore a great opportunity for CVA and the Valle d'Aosta Region where the Company operates.

In 2008 the climate and energy package established the binding target of 20% share of renewable energy in gross final energy consumption at EU level. The EU climate and energy framework in 2018 has identified a more ambitious binding target of at least 32% share for renewable energy in final energy consumption.

Since 2000, progress in technology has accelerated the deployment of solar and wind power, which can now compete with hydroelectric power in the energy market.

In keeping with regulatory and technological developments, in 2009 the CVA Group started expanding with a view to diversifying its business model and energy sources, as well as benefitting from certain incentives being introduced by the Italian government for the production of green energy. The goal was achieved primarily through acquisitions of shareholdings in companies which owned and operated solar and wind power plants. To a lesser extent, the Group also carried out and launched its own greenfield solar power projects (Figs. 4 and 5).

Over about 15 years the Company has developed a thorough value chain in the energy field from production to sale and distribution. To such extent, CVA is a jewelry box for a customer-centered experience: by opening the lid, our customers find the emeralds of pure green energy and service total reliability from sale to distribution. For those living in the Region the provision is also incarnated through 4 local shops run by CVA employees.

In 2018 CVA Trading delivered nearly 5 billion kWh of which 85.6 million kWh to regulated market clients and 4.8 billion to business and retail clients, for a total number of 147,054 points of delivery (POD) over the national territory. In 2018 CVA Trading was classified as the 11th energy sale operator in Italy with 4.9 TWh of energy sold at the national level. Figure 6 well represents the Company's evolution.

At the end of 2018 the approval of the Clean Energy Package with its eight legislative measures, provided ambitious targets for member states and a more articulate framework for the implementation of liberalised energy markets.

Fig. 4 Piansano Wind Farm. *Source* CVA

Fig. 5 La Tour Solar Plants. *Source* CVA

Accordingly, the Company aims at diversifying production as well as moving beyond regional borders. In 2018, CVA was the 4th producer from hydroelectric source and the 6th producer from renewable sources in Italy, settling among the major Italian players in the energy market (Tables 1 and 2).

3 From a Regional Identity Towards the Sustainable Development Goals of the 2030 Agenda

Over the last years the international scenario in the energy field has consistently changed. This paragraph will discuss how such changes impact on the business strategy of a local company such as CVA, positively bridging the gap from a regional to a nation-wide perspective.

The Energy Transition paradigm quickly spread over the last decade, accelerating its pace from the Paris Agreement in 2015 until the nowadays "Greta era". In 2015 about 200 countries stipulated the Paris Agreement which set the target to keep the global temperature rise well-below 2 °C above preindustrial levels [2]. All United Nations Member States adopted the 2030 Agenda on Sustainable Development Goals (SDGs) [3]. The 17 goals thematic issues ask the business

Fig. 6 CVA Timeline 2018

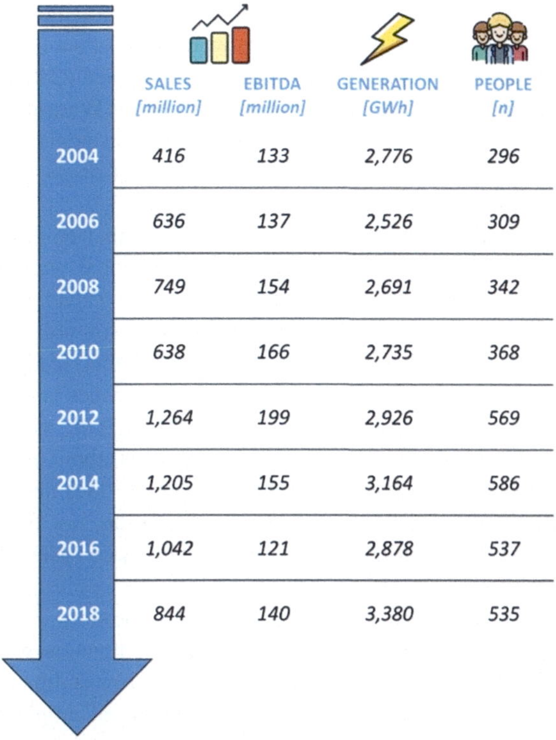

Table 1 CVA numbers: generation year 2018

+3 billion kWh generation from renewable energy sources (RES)		
HYDRO	WIND	SOLAR
6 dams for hydroelectric generation	8 wind plants in Aosta Valley, Tuscany, Puglia, Lazio, Campania	3 solar plants
32 hydroelectric plants in Aosta Valley 934.5 MW capacity 2.9 billion kWh generated	320 million kWh generated	16 million kWh generated
129 million m^3 the total dams' capacity: the equivalent of half the water consumed in Italy on a daily basis		
~€845 billion the value of production		

Source CVA

Table 2 CVA numbers: distribution year 2018

Deval is CVA's DSO		
4100 km the length of the distribution grid	130,000 POD	11° among Italian DSOs

Source CVA

world for thorough actions to support a holistic sustainability framework. Three years later, in 2018, the Intergovernmental Panel on Climate Change (IPCC)—the United Nations body for assessing the science related to climate change—has called for urgent action with its report on Global Warming by pointing out the failure of current trends to meet the Paris Agreement goals [4]. The consequences of global warming are such that by 2100 the impact on society and business as a whole can be seriously alarming.

World-wide, the search for solutions that could deliver the reduction of emissions between now and 2030 to limit global warming to 2 °C is spreading. The World Business Council for Sustainable Development (WBCSD)[1] suggests the concept of "business climate resilience" for policymakers and companies within the global climate agenda [5–7]. The other leading word in the framework of climate change is adaptation. Climate resilience is the capacity to develop responses to climate changes in order to both manage the associated business risks and seize new growth opportunities. Adaptation is about moderating harm and exploiting opportunities. The message is quite clear: "business as usual" cannot any more be the strategy to cope with climate transition, not from an environment perspective nor from a business point of view. This is consistent regardless of the nature of the business.

The legal framework supported this notion in 2014, with the EU Directive 2014/95/EU requiring that public interest companies with more than 500 employees disclose non-financial information, notably company policies regarding the environment, social responsibility, treatments of employees, respect of human rights, anti-corruption and bribery, and diversity on company boards. In Italy the decree 254/2016 has established such requirement as mandatory effective January 2017.

Last but not least, the Clean Energy Package [8], whose final delivery has been in 2018, supports the Energy Transition in a number of ways, providing a central role for high pace in RES deployment.[2] Accordingly, Italy has recently proposed its Integrated National Energy and Climate Plan (PNIEC) [9], detailing the vision for the five dimensions of the Energy Union: decarbonisation, circular economy, efficiency, and a rational and fair use of natural resources. The recently approved Ministerial Decree for RES deployment clearly highlights the need for integrating public incentives with private investments, in order to pursue the ambitious goals of Energy Transition.

In this broader scenario, the industrial strategy of CVA needs to tackle a few strategic questions: How can the company build on its natural hydroelectric heritage

[1]The WBCSD is a global organization of over 200 leading businesses working together to accelerate the transition to a sustainable economy.

[2]The Clean energy for all Europeans package consists of eight legislative acts which aim at facilitating the transition from fossil fuels towards cleaner energy and to deliver on the EU's Paris Agreement. The recast renewable energy directive has set a binding target of 32% for renewable energy sources in the EU's energy mix by 2030. The proposed electricity market design aims at integrating a greater share of renewable sources, with a focus on enhanced flexibility in the system and a more active role for consumers.

in order to promote environmental-friendly energy solutions and, at the same time, participate in the industrial mainstream towards climate change? Can the company seize an opportunity to harmonize the local community perspective with national developments, taking part in the world-wide change of society and business paradigm?

In 2019 the company has published its first sustainability report, although this is not a mandatory requirement for CVA. The reporting process enhanced the Group's positive exposure about its RES endowment, whose core business is contributing to environment protection and Energy Transition. On the other hand, there is indeed more room for sustainability practices and for pursuing a comprehensive strategy towards Agenda 2030. The sustainability materiality topics[3] identified by the Company stakeholders belong to four main fields in which CVA is investing in various ways: climate change, community, innovation and people.

Seeking the *right* balance between business goals, the National Climate and Energy Plan (PNIEC), and the Aosta Valley regional Energy Transition policy, CVA contributes to decarbonisation strategies and innovation with a number of initiatives. CVA generation from 100% RES accounts for saving 1.4 million tons of CO_2 (590 TEP). According to a recent study by TEHA and Enel, there could be a highly relevant contribution of the electric vector to decarbonisation, given that the energy mix contains a significant share of RES, reducing pollutant emissions on the one hand and contributing to introduce flexible energy systems for consumers on the other hand [2] (Fig. 7).

In October 2019 an incentive-based mobility law to promote the deployment of sustainable electric mobility in the Valley has been promulgated. Joining this perspective, a recent partnership with the company Becharge will allow CVA to install 250 electric charging stations all over the Region. The first of such charging stations has been symbolically placed at the foot of Mont Blanc, where the CVA Group also participates in the Save the Glaciers initiative, aimed at preserving the glaciers whose existence is seriously at danger due to climate change.

Furthermore, since 2004 CVA participates in the regional Glaciers Control Room whose mission is the safeguard of this natural heritage by monitoring and documenting changes in the glaciers' structure.

Within the innovation department, the company is currently collaborating with two university-based pilot projects on the development of local and RES communities. Such experimental projects focus both on the value of RES and on the central role of consumers in the Energy Transition. The EU Renewable Directive in

[3]Borrowed from the accounting field, the concept of materiality in sustainability reporting highlights those issues important for all the Company's stakeholders: "*Material topics for a reporting organization should include those topics that have a direct or indirect impact on an organization's ability to create, preserve or erode economic, environmental and social value for itself, its stakeholders and society at large.*" [10].

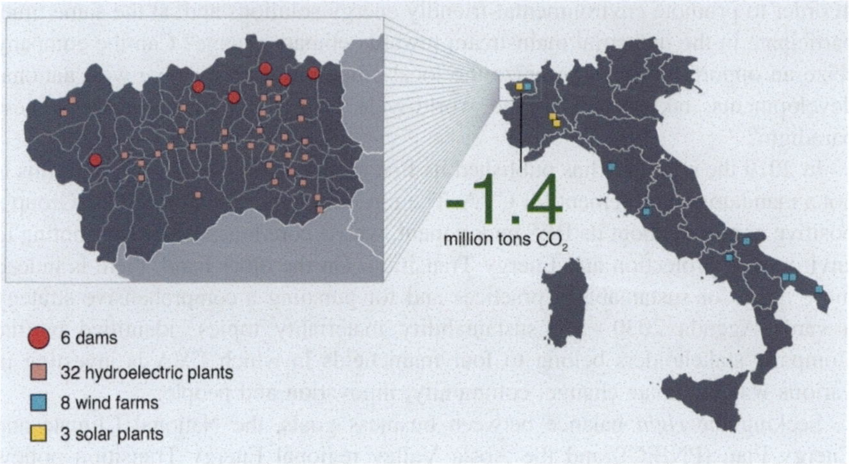

Fig. 7 CVA fleet. *Source* CVA

2018 specifically focussed on empowering consumers in producing their own energy. We learn from a research by the Milan Polytechnic [11], that it would be possible to deploy 450,000 energy communities all over Italy, 80% of them could be residential.

Such initiatives are of remarkable value, as their roots are in the Aosta Valley while their ambition is meeting the European and national goals towards Energy Transition. This is so, that the intertwining with developments at community level finds a natural ally in the regional government's plan towards a carbon-free Aosta Valley.

Certainly, Energy Transition is a responsibility for the business world in general and for the energy generation sector in particular. At the same time, sound government policies from State to Regions carry the duty to positively affect the outcome of Energy Transition, accompanying and coordinating the opportunities of such a massive economy transformation.

We would like to conclude by borrowing the words by Michal Kurtyka, President of COP24: "*Climate change and depletion of the world resources will affect countries and regions in a different time, and different ways. But it is a common cause and common responsibility*".[4]

References

1. 7th Environment Action Programme (EAP), European Commission, 2013

[4]Just E-volution 2030, p. 20, THEA, Fondazione Centro Studi Enel [12].

2. The Paris Agreement, FCCC/CP/2015/L.9/Rev.1, United Nations, December 2015
3. Transforming our World: The 2030 Agenda for Sustainable Development, A/RES/70/1, United Nations, 2015
4. IPCC Special Report on 1.5 °C, United Nation, 2018
5. Natural climate solutions: the business perspective, WBCSD, September 2019
6. Policy Enablers to accelerate the circular economy, WBCSD, August 2019
7. Business Climate Resilience Thriving Through the Transformation, WBCSD, September 2019
8. Clean Energy for all Europeans, European Commission, Publication Office of the European Commission, March 2019
9. National Plan for Energy and Climate (PNIEC), MISE, 2019
10. Technical Protocol: Applying the Report Content Principles. GRI, The Netherlands. Global Reporting Initiative (GRI). 2011
11. Smart Grid Report; Energy & Strategy Group Milan Polytechnic University; 2014
12. Just E-volution 2030. The socio-economic impacts of Energy Transition in Europe. The European House Ambrosetti, Fondazione Centro Studi Enel, 2019

Egea: The National Multi-Utility, Alba-Based

PierPaolo Carini and Rosario Bisbiglia

Abstract Originally established in 1956, Egea has evolved to become a true multi-utility company, with business ranging from energy (sale of power and gas, district heating, gas distribution network, renewable power production, public lighting) to environmental services (integrated water cycle, industrial and civil waste collection and treatment). Thanks to its unique structure, Egea is ideally positioned to support local governments in the management of public utility services. Egea's purpose is to bridge the gap between global markets and local communities by deploying new technologies that improve the lives of people (*glocal*). Egea believes that concrete efforts to achieve a low-carbon economy and implement a specific strategy for energy and environmental sustainability are opportunities for everyone: governments, energy companies, industrial companies, individuals. Egea is committed to investing and to helping to build a model for sustainable growth, focused on clean energy that can make people's lives simpler, in full respect of the territory not only in the North-West province but also in other Italian provinces.

Keywords Multi-utility company · Glocal · Public administration · Government · Industrial companies · Citizens · Energy · Environment sustainability · Carbon Free

1 Egea History: From a Gas Distribution Consortium to a Local Multi-Utility

Established in 1956, the initial focus of Egea was the distribution of natural gas in the city of Alba (CN). However, Egea's true beginnings can be dated to 1983 when, at the age of 61, Emanuele Carini decided to become an entrepreneur after a career as a top manager, first for one of Italy's most important national gas operators, and then for a large municipal multi-utility. With the arrival of Carini, Egea expanded its geographical coverage to over 40 surrounding cities.

A. Gilardoni (ed.), *The Italian Utilities Industry*,
https://doi.org/10.1007/978-3-030-37677-2_11

In 1986, Egea started diversifying from natural gas distribution with its first project in district heating in the city of Alba. This project was followed by 33 similar initiatives in other cities (Fig. 1). Diversification into related businesses continued in the '90s with the beginning of the integrated water cycle (1991) and waste collection (1994) services. During this period, Emanuele Carini's son, PierPaolo, decided to move to Alba, ending his academic career, to join Egea.

In the early 2000s, with the liberalization of the gas and power markets, Egea decided to become a national player in the energy sector. From the city of Alba, Egea began its expansion into the rest of the Cuneo province and then into other areas of North-West Italy, becoming a reference in the regions of Piedmont, Lombardy, and Liguria. This was also made possible thanks to the entry of new private investors, representatives of Alba's manufacturing community that believed in Egea's growth project. Relationships between Egea and key local players are now being intensified even further, as is demonstrated by the "AlbaPower" project (see Fig. 2), set up in partnership with Ferrero, for the construction of a cogeneration plant that supplies the energy needed by the confectionery manufacturer and produces heat for the city.

The successful experience in district heating pushed Egea to collaborate on the design, development and management of new power stations and heating networks in several other small cities, including Fossano, Acqui Terme, Carmagnola, Piossasco, and Cairo Montenotte. In addition to district heating, new renewable energy activities in photovoltaic systems were launched by Egea in Fossano, Turin, Valenza and Enna. To these, we must add several biogas plants, including Ozegna

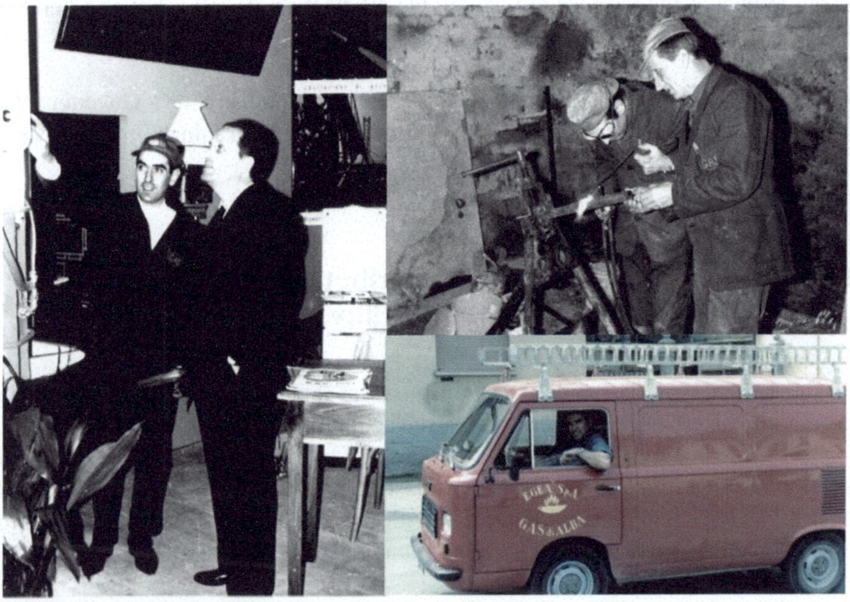

Fig. 1 Some pictures from Egea's early years. *Source* Egea

Fig. 2 The "AlbaPower" cogeneration plant built by Egea together with confectionery and chocolate manufacturer Ferrero. The plant supplies the energy needed by the industrial site and the district heating for the city of Alba

Fig. 3 The "Tanaro Power" hydroelectric power plant in the city of Santa Vittoria d'Alba

and Marene: the former is powered by livestock waste and plant biomass, while the latter is mainly fueled by cattle waste.

Egea confirmed its interest in renewable energy plants when it started a project in collaboration with Fiat Chrysler Automobiles and "CNH Industrial" on the production of biomethane for the automotive sector, in order to reduce the consumption of fossil fuels.

These positive experiences gained over the years allowed Egea to face a new challenge with the construction of "Tanaro Power" in the city of Santa Vittoria d'Alba (see Fig. 3), the Group's first hydroelectric power plant. Even in this case, special attention was given to finding innovative solutions that minimize environmental impact.

Today, Egea has reached a consolidated turnover of close to €1 billion and is widely considered a true *multi-utility company*, with business ranging from energy (sale of power and gas, district heating, gas distribution network, renewable power production, public lighting) to environmental services (integrated water cycle, industrial and civil waste collection and treatment). As a result, the initial acronym of Egea as "Esercizi Gas E Affini" was subsequently changed to become "Ente Gestione Energia e Ambiente" (Figs. 4, 5 and 6).

2 Egea's Unique Model: A Partnership Between Public and Private Stakeholders

In 1997, Egea changed its ownership structure by opening up equity to all the municipalities in which the Company was supplying services. In the following years, new shareholders joined Egea, mainly from the local industrial and financial communities.

As of the end of 2018, the Carini family owns 59.75% of Egea SpA shares, while private shareholders control 31.12% and public shareholders 9.13%.[1]

Egea's shareholding structure is comprised of a large number and wide variety of both public and private entities. These are mainly concentrated in the North-West of Italy: public entities ranging from small municipalities of a few thousand inhabitants to far larger ones, and private shareholders ranging from large multinational enterprises to small local firms or private investors. Egea is building on this collaborative characteristic even further, not only in the original territory, but also in several other areas of Italy, which can be defined as "rural". This is a new, innovative *glocal* development model; we at Egea call "federal aggregations" (i.e. aggregazioni federative).

This approach makes Egea unique in the business arena in Italy; it has become even more important in a period of scarce resources for Italian public administrations. Benefiting from its multi-utility approach, Egea is uniquely positioned to support local governments in the management of public services. In other words, Egea can collect, synthesize, elaborate (and hopefully design and solve) many of the various issues and opportunities coming from local stakeholders. In a historical period where the community is affected by social malaise on both local and global levels, this ability of Egea is even more important, especially when it comes to the environment and the rational use of energy (Fig. 7).

It is important to underline that Egea is not intending to replace local governments or other local stakeholders but rather to act in support of the development of

[1]Egea CSR Report, 2018.

Fig. 4 Grinzane Cavour Castle (CN), a symbol for the winegrowing areas of Langhe, Roero and Monferrato in Piedmont and also a UNESCO World Heritage Site. *Source* Egea

Fig. 5 The beach at Andora (SV), on Liguria's Western Riviera: some of the areas in which Egea is now firmly established. *Source* Egea

the local community. Egea is supporting the Country while working in harmony with public administrations and local industrial systems.

In this respect, sustainable development in coordination with communities may present not only a genuine industrial recovery but also an opportunity to relaunch positive human values among every member of the local community, such as municipalities, professional associations, and industrial companies. Luckily, the world is changing, and as a result the way of doing business has to adapt too. Strengthening local communities is a shared objective of stakeholders with different competencies, roles and capabilities, all strongly committed to the protection and development of the local community (Fig. 8).

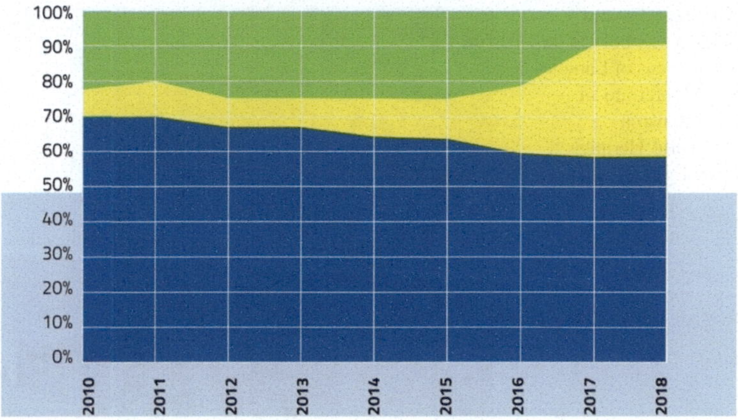

Fig. 6 Historical evolution of Egea's shareholding structure. The Carini family stake is shown in blue; private shareholders in yellow; public shareholders in green. *Source* Egea

3 The Organizational Model of the *Glocal* Multi-Utility

It is no coincidence that Egea was founded and developed in one of the most excellent Italian "suburbs": the Alba area, characterized by a strong entrepreneurial spirit, by the ability of creating partnerships and cooperation between public and private entities, by the highest quality of life and services, and where landscape, nature and biodiversity protection become a further element of valorization and production of wealth and excellence. Born in such a context, Egea was then developed by inflecting work and ethical awareness into economic, environmental and social sustainability strategies. It has become an example, an entrepreneurial model capable of putting efficiency and creation of value for the territory and its inhabitants together.

This style has been clear since Egea's foundation. The Group evolution was based on two principles: closeness to the territory and industrial effectiveness. Egea's development as a multi-utility company has evolved along these two concepts: new areas of business, driven by the same green economy philosophy, have grown alongside the historical sectors where the Group companies are leaders. District heating, public lighting, energy production from renewable sources, biogas and biomethane, electric and sustainable mobility are just some examples of an increasingly expanding range, designed to promote local well-being for both private customers and public administrations.

Egea has proven to be naturally versatile in different sectors pertaining to public utilities, energy and environment.

Multi-utility is in our DNA, it is what defines Egea.

This is why, over the years, the Group has branched off into a series of companies and special-purpose enterprises, often in partnership with leading players in the entrepreneurial world, related to small areas and capable of meeting specific

Fig. 7 Example of advertising created by Egea to promote its commitment to the development of the local community. *Source* Egea

needs through targeted projects. In fact, Egea Group's organization (see Fig. 9) has been developing in an increasingly structured way: the geographic expansion towards territories interested in Egea's services has led to new partnerships, cooperation and acquisitions of shareholdings in other enterprises, which strengthened the Company's growth, by enhancing existing landmarks or building shared improvement paths.

Egea's corporate model is designed to extend its growth potential, thanks to the holding's central function, which carries out policy and control over the other operating companies, while ensuring a reduction of the investment risk for shareholders. By investing in Egea, shareholders can participate in a plurality of

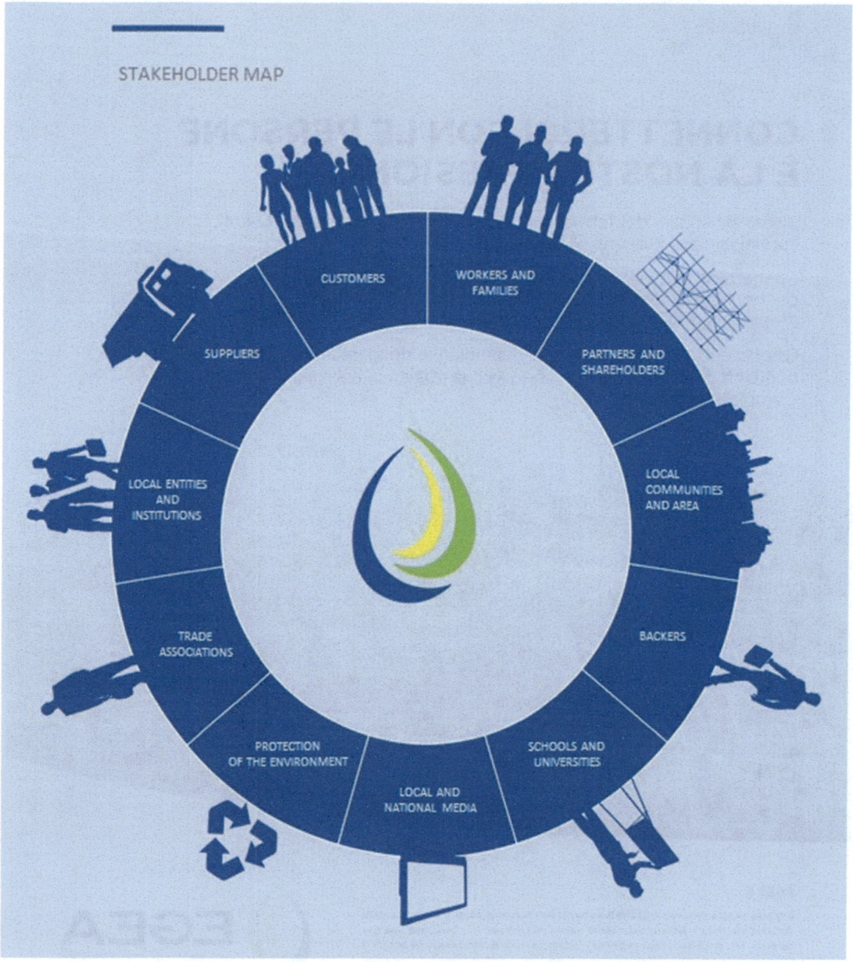

Fig. 8 Egea's map of stakeholders. *Source* Egea

businesses, all pertaining to the public utilities, energy and environment fields. Moreover, the holding, by ensuring a centralized management of certain activities shared by all subsidiaries, obtains better results from an effectiveness and efficiency standpoint, compared to what each single company could do. The purpose of this holding is to create value for its shareholders and, most of all, for all its associates and subsidiaries. Value creation is not limited to the Company: it affects the entire territory, thanks to the varied and multiform composition of Egea's equity. This model best expresses itself in the Italian suburban territories, where the need to collaborate and cooperate to overcome the challenges related to small companies and to the lack of infrastructure is greater.

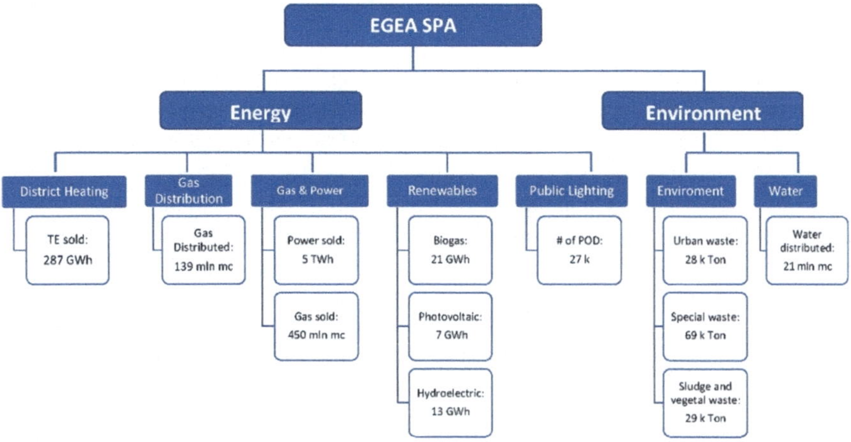

Fig. 9 Egea's organizational model and key business figures

With each passing day, Egea becomes a multi-utility group that meets varied local needs and can offer "tailored-made" solutions for Italian "suburban" areas: this is the formula through which the multi-utility Group has become more and more rooted in the entire Italian North-West, with great growth potential in other Italian areas.

This way, Egea has found its reference territory in those "suburban" areas, where the majority of Italians lives.

4 Egea as a Driving Force for Development

Technological progress is about improvements regarding the way in which goods and services are manufactured, marketed and sold. The level of technology in a country depends both on the extent to which it is exposed to the international development of technologies and its capacity to absorb them. While in big cities like Milan, it is easy to be exposed to technological progress, or even to be the home of trials for new services, in small cities, especially those in rural provincial areas, the risk of technological divide is high.

The role of Egea has been—and will be even more so in the future—to bridge the gap between global markets and local communities, by bringing new technologies that improve people's daily lives. Egea is a good example of the global to local—better known as *glocal*—model. In fact, Egea is doing this by leveraging its distinctive characteristic:

- being the smallest player among the biggest ones makes Egea able to understand the needs of local communities;
- being the biggest player among the smallest ones makes Egea able to take advantage of economies of scale in providing best-in-class services to local communities.

For example, Egea was selected by EnelX[2] to be the key partner in the development of the network of electric vehicle charging stations in 85 towns in the regions of Piedmont and Liguria, mainly focusing on small villages. The partnership between the two operators will promote e-mobility in a territory characterized by a thriving tourist industry for both foreign and domestic visitors. The plan is to install 255 charging points by 2022 without any cost to the 85 towns.

Today, Egea acts as a point of reference for its stakeholders, such as local municipalities, regional governments, private enterprises and citizens, all of them committed in the development and wellbeing of the local community. Based on this model, Egea is leading the study, processing and proposal of new policies that are subsequently deployed at local levels. Egea takes the responsibility of financing large projects with positive environmental impacts; it takes upon itself the task of sourcing power and gas on wholesale markets at competitive prices in the interest of its clients, and also takes on responsibility to look for and select best-in-class technologies to reduce energy consumption in industrial processes or residential buildings. For example, Egea considers the development of district heating in the city of Alessandria as the most important case of environmental and energy efficiency (EE) in urban areas. This project, with a total investment in excess of €90 million will not only reduce pollution in Alessandria but also generate positive externalities, since most of the capital expenditure is made in favor of local enterprises.

Ultimately, the deployment is not just carried out by Egea autonomously, but in partnership with its stakeholders, each of them characterized by own peculiarities, needs and knowledge of the local community. In this way the deployment is strongly connected with the local reality. Consequently, global joins local and becomes *glocal.*

5 Sustainability and the Environment: Key Elements for a Unitary Territorial Policy and for the Relationship with Communities

The global balance of the planet is increasingly linked to local actions at national, regional and local level. Concrete efforts to achieve a low-carbon economy and implement a specific strategy for energy and environmental sustainability are an opportunity for all stakeholders: governments; energy companies; industrial companies; individuals, etc.

At present, the private sector has an ethical duty and a priority role in the fight against climate change. By implementing emission reduction initiatives in line with global targets, such as the Paris Climate Agreement for reducing greenhouse gas

[2]EnelX is a company of the Enel Group dedicated to diversified businesses; see in this book the chapter regarding Enel.

emissions and the 2030 targets of the European Union, the private sector can turn a threat into an opportunity.

The adoption of all possible decarbonisation measures and the increase in the sustainability of our society require the implementation of projects aimed at optimising available resources, as well as the use of innovative management approaches. The ways in which we live, think and produce must gain greater focus on ethics, which must be translated into awareness-raising actions and the formation of a new class of technical and local government. Doing something concrete against climate change is now necessary and indispensable for our own lives as well as for future generations.

This was clear for Egea when it planned and developed photovoltaic plants in abandoned quarries, or biogas plants that reduce the unpleasant odours from pig farming. Today, with regard to these objectives, Egea in partnership with public administration, Confindustria Cuneo, GREEN[3] of Bocconi University and the Polytechnic of Turin-DENERG, has decided to launch the *Granda Carbon Free Objective* project.

While there are several examples of top-down approaches to implement sustainability actions (e.g. the Siena Carbon Free Project, which began in 2006 and allowed the Province of Siena to achieve "carbon neutrality" in 2011), the *Granda Carbon Free Objective* project aims at using a bottom-up approach. In accordance with national and regional energy strategies, the starting point will be accurate knowledge of the needs and specific aspects of the territory.

Starting from this knowledge, pilot projects can be identified, involving different actors within the territory: private companies, energy professionals (designers and producers), energy service companies, credit institutions and/or other financial actors and, finally, local governments focused on territory development. Through feasibility studies on these pilot projects, concrete actions will be identified to achieve a substantial reduction in atmospheric emissions and an increase in sustainability. These actions will address the issue of reducing consumption and developing innovative resource management solutions, in order to create synergies among stakeholders and optimise the use of available resources (Fig. 10).

One possible area to study innovative services in the field of sustainability is Alba and the Langhe, whose wine landscapes have been recognized as a World Heritage Site by UNESCO. This recognition has increased the tourist appeal of the area, making it essential to adopt environmental sustainability solutions. Additionally, the socio-economic landscape is characterised by the presence of both large companies (e.g. the Ferrero and the Miroglio groups) and small businesses (especially in the agricultural and tourism sectors). The size of these entities often does not allow them to make major investments in environmental sustainability, and therefore, aggregation and the creation of synergies is essential to achieve these objectives. The results of these pilot projects can be extended and reapplied across

[3]The Centre for Research in Geography, Resources, Environment, Energy and Networks.

Fig. 10 Picture of the Langhe area. *Source* Egea

the entire territory of the province of Cuneo, creating the conditions for the definition of local strategic policies by the public administrations in charge.

Technical skills and knowledge of the territory are essential ingredients for a successful decarbonisation project. The involvement of the local population through presentations of old and new initiatives, together with the results obtained is therefore crucial to ensure the replicability of the approach and the long-term success of the identified solutions.

6 Going Forward: New Development Perspectives

For a multi-utility like Egea there are a few strategic elements that one must not lose sight of: customer centrality and satisfaction; continued reduction in environmental impacts, and the search for innovative technologies, shared with the territory, to improve the quality of networks and services. As far as customers are concerned, Egea must keep building more and more innovative and personalised proposals for each individual customer. This is the reason why Egea will continue to invest in new stores, in trained and problem-solving-oriented staff, capable of maintaining an ongoing dialogue with customers, in smart technologies, and in improving its internal organisation to ensure efficiency, the ability to prevent problems and, when these occur, to provide responses that are better than those of the average competitor.

Furthermore, actions will be pursued in the modernisation and the efficiency of existing networks and facilities. Thanks to successful collaborations with the academic world, Egea will continue to seek out new technologies, aimed at delivering smart services that improve the quality of life for people and their territories.

Egea's future strategy will remain focused on extremely functional infrastructures and technologically advanced projects that focus on EE, the development of renewable energy, support for sustainable mobility, and the deployment of services for people which are traceable, efficient and economical.

The excellent results achieved in our public lighting business cannot be considered an end point. They are a first step towards even more exciting challenges, to be tackled on the basis of all projects achieved so far and to be used in order to offer local governments innovative infrastructure solutions, combining EE with advanced sensors.

Our green mobility initiatives are now starting to reap the benefits of all the efforts made in the past years, with the progressive installation of charging stations for electric vehicles and with the production and distribution of biomethane, both aimed at replacing traditional fossil fuels.

Egea's leading role in district heating will be consolidated with the implementation of the Alessandria project. Once more, this demonstrates how Egea's model works well, bringing together technology, entrepreneurial intuition, improved environmental performance and shared economic benefits with local stakeholders.

Going forward, Egea wants to continue to invest and to help build a model of sustainable growth, focused on clean energy that can make people's lives simpler, in full respect of the territory not only in the North-West province but also in other provinces of Italy. As such, Egea is committed to expanding its unique *glocal* model in support of other territories across the entire Country. That is, Egea is looking at the next few years as a very important opportunity to play the role of protagonist in several different areas of Italy as a *glocal* player, particularly in the optimisation and development of real and deep connections between public administrations, industries, and investors. We are confident that this will bring big results for our environment, our economy, and our community.

Part IV
Network Operators: A Changing Role and Challenges for the Future

Terna: Challenges for the TSO. From Liberalisation to Energy Transition

Luigi Ferraris

Abstract Terna, the Italian electricity TSO, has experienced since its foundation many changes in its role. Today the company is moving towards a new phase in this evolutionary path. The energy landscape is experiencing a period of deep transformation: the Energy Transition, with the growing amount of intermittent renewables and distributed sources connected to the grid, puts significant pressure on network infrastructure. At the same time, the concern for environment protection is increasingly important in the public debate, making the decision for new development projects more delicate than ever. Terna is actively responding to these challenges: on one hand it has launched an ambitious investment plan to increase the resilience and stability of the national electric system to allow the decarbonisation of the power sector, while on the other it has set up a new model of stakeholders' engagement to guarantee that each new project is sustainable on a social and environmental ground.

Keywords Power transmission · Sustainable grid development · Stakeholders' engagement · Energy transition

1 Terna's Story

1.1 Liberalisation of the Energy Sector

Terna is the company responsible for transmission and dispatching activities on the high- and extra-high-voltage Italian electricity grid. With over 74,000 km of lines and 884 electrical substations, it is the largest independent transmission system operator in Europe and a global leader.

Terna guarantees the security, efficiency and sustainability of the grid, minimising overall costs for Italian families and businesses. In addition to these activities *(Italian, Regulated)*, Terna is developing businesses on the free market in Italy, leveraging on the technical capabilities of the core business alongside innovation *(Unregulated)* and is investing in countries, with a stable political and

A. Gilardoni (ed.), *The Italian Utilities Industry*,
https://doi.org/10.1007/978-3-030-37677-2_12

regulatory landscape, which are in need of a modern electricity infrastructure, also through the collaboration with energy operators with sound foreign business *(International)*.

Terna manages all of its activities with a strong focus on possible economic, social and environmental impacts and adopts a sustainable approach to its business in order to create, maintain and consolidate a relationship of mutual trust with its stakeholders, supporting the creation of shared value.

Terna's story began with Italian Law 1943 in the year 1962, which assigned Enel all phases of the previously privatised electricity sector, marking its nationalisation. This state monopoly allowed the electrification of the Country, but it was the process of liberalisation, promoted by the European Union more than thirty years later and aimed at independent grid management, the true driver behind the current operating environment.

The implementation of Italian Legislative Decree 79/1999 (the "Bersani Decree") provided for the liberalisation of the electricity sector and, among other things, the separation in the ownership of the NTG (National Transmission Grid) from its operation (transmission and dispatching activity), according to the Independent System Operator model. Accordingly, two companies were established: Terna, owner of the Italian national transmission grid, and GRTN (Gestore della Rete di Trasmissione Nazionale—Italian National Transmission Grid Operator).

1.2 Terna and GRTN

On 1 May 1999, following the unbundling process provided for in the decree, the company Terna was set up within the Enel Group. According to this Legislative Decree, "Terna's activities regard running and maintaining the Enel Group's National Transmission Grid systems, as well as developing them, according to guidelines issued by GRTN (Gestore della Rete di Trasmissione Nazionale)".

The subsequent Italian Prime Ministerial Decree of 11 May 2004 sanctioned the consolidation of ownership and operation of the National Transmission Grid under Terna, which was listed on the stock exchange on 23 June 2004. From 1 November 2005, it assumed responsibility for transmission and dispatching of electricity on the high- and extra-high-voltage grid all over the Country, acquiring the residual portions of the grid from other operators. The majority shareholder of Terna is CDP (Cassa Depositi e Prestiti), with a 29.99% stake in the capital.

At the same time, GRTN changed its name to GSE (Gestore dei Servizi Elettrici —electricity services operator) and began to focus on promoting and encouraging electricity generation from renewables.

1.3 Renewed Investment in Electricity Infrastructure

The process of liberalisation of the electricity market saw the relaunch of investment in production capacity and grids, which had been stagnating for several years. In 1999, the total installed power of Italian generation facilities was around 76 GW. Through the liberalisation process, generation facilities had reached around 105 GW in ten years, with a solid 22 GW from new thermoelectric plants. This period of new investments in thermoelectric plants (mainly gas-fired), boosted also by legislative measures such as the "Power Station Unblocking Decree" of 2002, gave the Country a completely renewed and extremely efficient generation fleet. In subsequent years, following the 2008 economic crisis and the related drop in electricity demand, the Country went through a period of overgeneration, where installed power regularly exceeded peak output. At the same time, development of renewable energy sources (RES) began and renewable installed power grew exponentially, rising from around 4 GW of wind and photovoltaic capacity in 2008 to over 30 GW in 2018. This growth, combined with the stagnation of electricity demand, reduced profits for conventional thermoelectric plants, launching a phase of "mothballing" and closure of a significant portion of thermoelectric facilities, which dropped from around 76 GW in 2013 to 61 GW in 2018.

The energy system is currently going through a huge transition, requiring proper management to ensure that the adequacy, security, resilience, quality of supply and economic efficiency of the electricity system are always guaranteed. This requires investment in grid infrastructure that is proportional to the one in renewables. Every euro spent on new renewable generation requires more than one euro of investment in electric infrastructure and related services.

With this in mind, Terna has presented its largest ever investment plan. In its 2019–2023 Strategic Plan, the Group has planned a total of €6.2 billion in investments to meet new electricity system requirements, integrating sustainability, dialogue with communities, development of expertise and promotion of innovation. Over €3 billion have been allocated to the development of the national transmission grid, of which more than €2 billion are devoted to upgrading and efficiency activities, primarily aimed at service quality improvement and electricity grid digitalisation. The other €1 billion will be invested in the Defence Plan for creating and installing devices to increase grid security and stability. In its investment, Terna adopts an approach targeted at listening communities and local populations and involving stakeholders from the start of the projects planning phase. This is a central component of the work process and ensures sustainability from a social, economic and environmental perspective. Transparent discussion and dialogue with local areas is a key aspect of Terna's growth strategy, hence a close relationship and interaction with communities represents a priority. Terna's service has its roots at local level: here it generates not only economic benefits, but social and environmental ones as well.

Fig. 1 Maintenance work on Terna's infrastructure. *Source* Terna

Since 2005, Terna has invested over €13 billion, now managing almost twice as many high- and extra-high-voltage lines as it did in June 2004 (which was around 39,000 km) (Fig. 1).

1.4 The New Challenges

Today, the energy system is going through a period of profound transformation, in a context of climate change and scarce resources compared to the needs of the population.

In just a few years, we have gone from a situation of overcapacity to one of increasing depletion of reserve margins, also due to the poor contribution from non-programmable renewables to system adequacy. The last 20 years have seen an increase from 800 to 800,000 generation points, which is expected to rise further towards one million. This means that the activities of securely and adequately operating the grid are today strongly affected. The electricity grid must be meshed and digitalised to be able to optimise the secure use of renewable resources integrated in the system, with the minimum possible energy dispersion (Fig. 2).

The challenging targets set out in the PNIEC (Piano Nazionale Integrato Energia e Clima, Italian National Integrated Energy and Climate Plan), which calls for the phase-out of coal plants by 2025 (around 8 GW) and the development of further renewable capacity by 2030 (around 40 GW), will exponentially increase critical issues around Electricity System management.

In an increasingly complex and constantly evolving energy context, innovation and digitalisation are two essential factors and represent two key pillars in Terna's strategy. We need to switch from a management approach based on watts to one

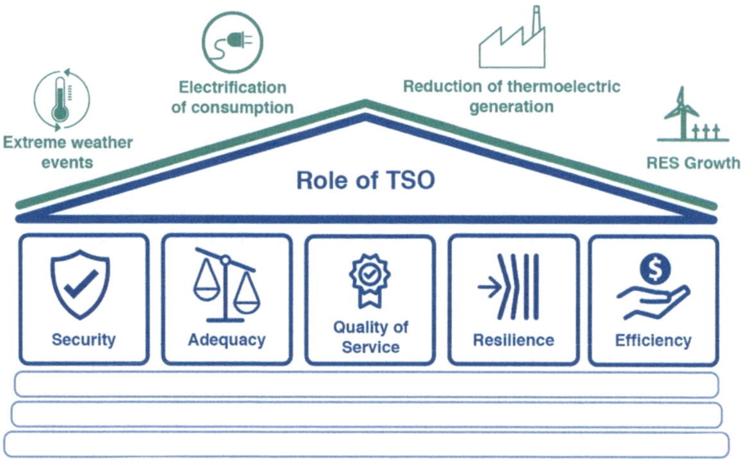

Fig. 2 The role of the electricity TSO in the challenging context of the energy transition. *Source* Terna

based on bytes, that is data management, which plays an increasingly important role in an ever-more interconnected grid, also at the European level.

2 Sustainable Grid Development

2.1 Terna's Approach

Terna's main activity corresponds to licence obligations and translates into a constant commitment to ensuring that the entire Country is guaranteed a secure, high-quality electricity service, with improved pricing through sustainable management and development of the transmission grid. The current transition towards a carbon-free economic system has placed Terna in a guiding and enabling role, aligned with the goal of supporting optimal integration of RES. Increasing use of RES and development of the electricity grid go hand in hand, as the latter is an enabling condition for the former.

The primary planning tool is represented by the National Transmission Grid Development Plan that Terna prepares every year. This document specifies projects planned for the next ten years, the state of progress of works planned in previous years and the investments required, defining at the same time the collective requirements for a secure and efficient electricity service and Terna's commitment to meet these.

Given these goals of general interest, all planned investments are subject to prior cost-benefit analysis, which expresses the comparison of the investment costs and benefits associated with implementation in monetary terms. A positive cost-benefit ratio is a prerequisite for inclusion of every single investment in the Development Plan.

The Development Plan is subject to evaluation and approval by the Italian Ministry of Economic Development, also following public consultation performed by the Italian Regulatory Authority for Energy, Networks and Environment (ARERA) and subject to strategic environmental assessment by the Italian Ministry of the Environment and Protection of Land and Sea (MATTM) alongside the Italian Ministry of Cultural Assets and Activities (MiBACT), with the purpose of integrating environmental considerations into the plan-creation process, guaranteeing sustainability.

The 2019 Development Plan is aligned with the evolution path of the European electricity sector towards progressive decarbonisation. It meets the challenge of an evolving climate scenario characterised by more frequent extreme weather events, placing ever-greater focus on the territory. The plan revolves around the following pillars.

Decarbonisation: the transition of the electricity system towards complete decarbonisation requires activation of all necessary levers for the full integration of renewable production in order to reduce emissions and guarantee system security.
Market efficiency: the structure and mix of European generation facilities, and particularly those in Italy, are undergoing major transformation, as is grid development, in line with new European Directives.
Service security: it is necessary to guarantee the security of the national electricity system while at the same time creating an increasingly resilient system capable of facing critical events external to the system.
System sustainability: this is understood as the ability to conceive, design and realise projects on the basis of strict analysis capable of maximising environmental and economic benefits.

2.2 *From Participatory Consultation to Participatory Design: Social Acceptance of Infrastructure*

The social acceptance of new infrastructure in Italy is one of the most delicate problems we face today, and inevitably it has technical and economic as well as ethical and social implications, particularly for works of public utility and social value. Collective benefits, which extend far beyond the boundaries of the areas affected by the works, cannot always be directly associated with the areas involved, while the inconvenience due to the works are site-specific and easy to identify.

Opposition from communities to new works is due to lack of confidence in institutions, frustration over infrastructure development lacking sustainable and integrated planning, the blinkered defence of one's own possessions and space (Not In My Back Yard or the NIMBY syndrome), the fear of the effects that the works will have on health and social cohesion, the inability to understand the technical aspects of projects and *a priori* non-acceptance of any change. This opposition is then capable of applying increasing pressure on local politics, social opinion and the bodies called upon to make decisions in the authorisation process, potentially obstructing work on construction sites once the project has been authorised.

It is therefore necessary not only to involve local authorities in the design process, but also to directly involve the local population, who, even if informed by administrators through open municipal meetings or other events organised during the consulting process, requires a central role. A new form of "participatory design" is needed: a major revolution of the process consisting in a new method of consulting founded not on negotiation, which is often late and ineffective, but on early-stage social design. Basically, what is required is a wider involvement during the planning phase, in terms of content and actors. This recognises the importance of social self-realisation, starting from the assumption that the social actors themselves are best informed on the social requirements of local areas.

ERPA Criteria

In local consulting, one of the most effective tools for selection of the least invasive options is represented by ERPA (Exclusion, Repulsion, Problem, and Attraction) location criteria.

The area to be analysed, with its land-use classifications and relative protection restrictions, is classified on the basis of criteria that express its level of suitability to host electricity infrastructure. Terna and regional authorities have agreed on a system of criteria based on the following four categories.

Exclusion: areas in which development is not permitted.
Repulsion: areas which should preferably be avoided, except in the absence of alternatives or when there are only less environmentally compatible alternatives.
Problem: areas problematic for objective reasons, documented by bodies involved, and which therefore require further land analysis.
Attraction: areas to be preferred where possible, subject to verification of the load capacity of the land.

The use of GIS (Geographic Information System) technology allows integrated consideration of all layers of information regarding different types of land use and protection restrictions (land, naturalistic, cultural, landscape, etc.), appropriately redistributed within the different ERPA categories in order to identify the most sustainable location for Terna's projects.

2.3 Stakeholder Engagement

Terna's stakeholder engagement model has been perfected over the years. Today, it is increasingly necessary to develop models of public consultation and trust-based cooperation with local authorities in order to agree on technological and design choices in advance, a method that allows a better understanding of social and environmental values of the local area. The comparison of different infrastructure development models highlights the path taken in fine-tuning stakeholder engagement at Terna. Throughout the four phases of the development models, from identification of an electricity requirement to project completion and commissioning (planning, consulting, authorisation and realisation), the positioning and level of stakeholder involvement changes (Fig. 3).

In the *Classic or Standard Model* during the planning phase, not all local families have access to the project and they are excluded also during the consulting phase, as the party putting forward the project prepares the design autonomously and submits it for authorisation (t3). This is the phase in which opposition may grow, generally structured in one or more groups of citizens, supported by environmental associations and local authorities that mobilise against the initiative. The risk is that the work will not be authorised or, if it is, that the site will have to be managed with the help of the police at the construction stage (Fig. 4).

In the *Communicative Model* of stakeholder engagement, dialogue with local bodies is launched, aimed at informing authorities of the need for infrastructure and about the project decisions already made. In this model, planning is not followed by consultation and the project is defined autonomously by the promoting parties. However, at times, the stakeholders oppose the work also at the construction site stage (Fig. 5).

In the *Consultative Model*, social stakeholders become aware of the project during the authorisation phase. With this approach, planning is followed by a consulting phase involving local bodies before final design. The consulting process ends with the signing of Memorandums of Understanding on the technical and

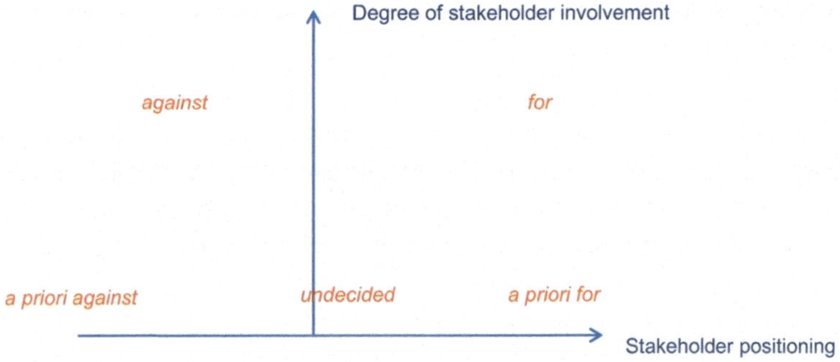

Fig. 3 Positioning and level of stakeholder involvement in relation to the different stakeholder engagement models. *Source* Terna

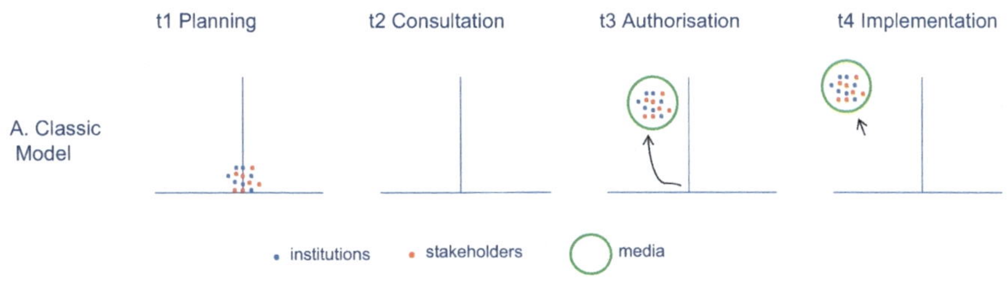

Fig. 4 Classic or standard model. *Source* Terna

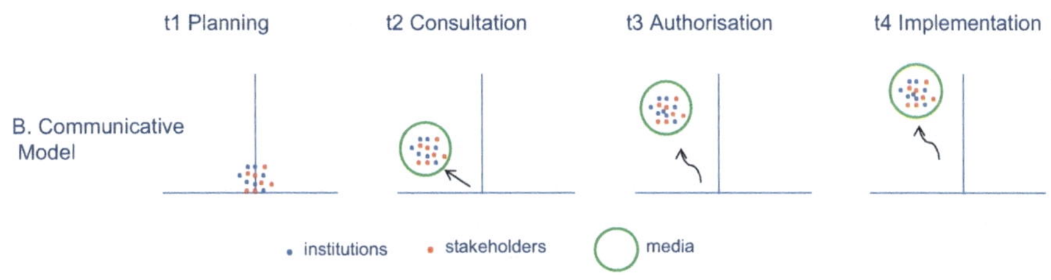

Fig. 5 Communicative model. *Source* Terna

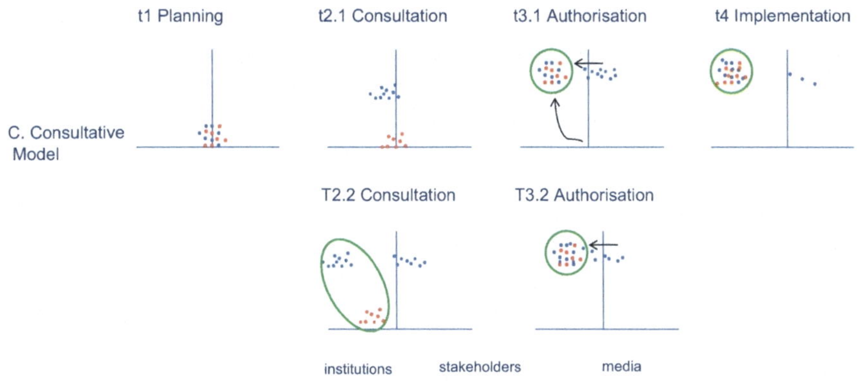

Fig. 6 Consultative model. *Source* Terna

location choices and those regarding grid rationalisation. In this context, social stakeholders are either marginally involved or not involved at all, becoming aware of the project during the authorisation phase, in which opposition is triggered not only against the proposing party, but often primarily against the authorities. Additionally, in this case, the project risks not receiving authorisation, although less so than in the previous scenarios. The construction phase remains problematic, albeit to a lower extent (Fig. 6).

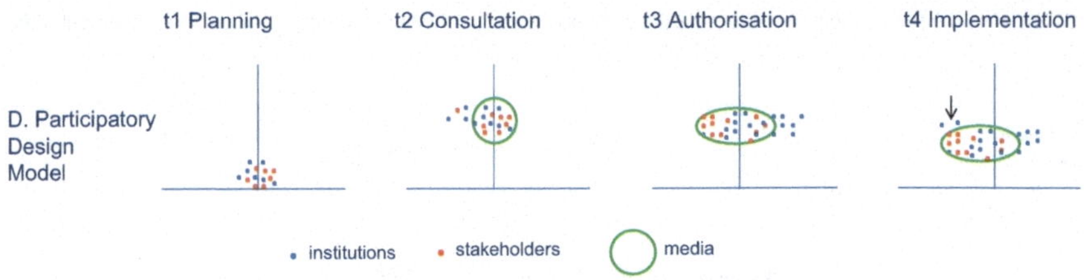

Fig. 7 Participatory design model. *Source* Terna

The *Participatory Design Model* (Fig. 7) is the approach currently adopted by Terna, and is considered the most advanced stakeholder engagement method to date. During initial phases, in addition to collaboration with local bodies, there is dialogue with social stakeholders involved, alongside technical meetings with municipal councils. The first meetings with citizens and the environmental and sector associations, through the open day tool (i.e. *Terna incontra*, see below) or dedicated events, are aimed at presenting the power system requirements and gathering information on the local area and the environment. This is followed by other meetings and open days aimed at assessing and agreeing on areas to be excluded from analysis of possible solutions. Further discussions and open days are dedicated to possible works locations resulting from previous dialogues and from the latest project optimisation indications. In this approach, rationalisation projects for existing electricity grids, as well as identification of possible compensatory actions, are fundamental for successful social acceptance of the initiative. This is not only the product of dialogue with local authorities, but with citizens and environmental associations, above all. Stakeholder awareness and participation is brought forward to the consulting phase, and opposition is also far weaker in the subsequent construction stages.

The *Large-Scale Design Model* (Fig. 8) is currently only theoretical and has not yet been tested in the field. The planning, consulting, authorisation and realisation phases are brought forward and partially conditioned by intense, prior analysis of the local context, with evaluation of requirements and opportunities for development of the various sectors and leveraging on available resources. Opposition dynamics may be completely different, and greater acceptance and participation may be seen, along with local and regional political support and positive media attention.

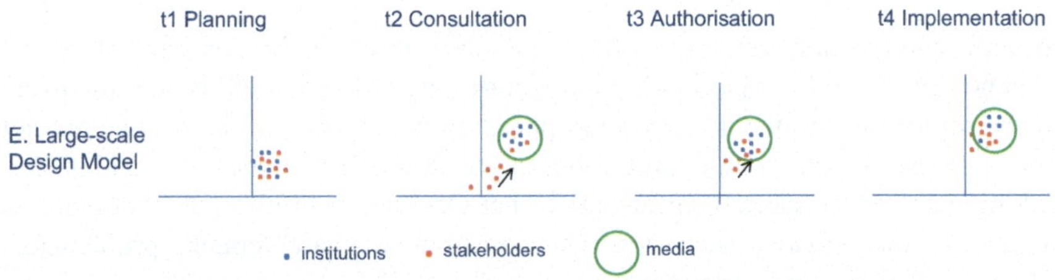

Fig. 8 Large-scale design model. *Source* Terna

2.4 Terna Incontra… *Open Days*

Early consultation and participation in the design process have been extended to include citizens through the *Terna incontra*… (literally: Terna meets…) open days. These are whole days dedicated to the local population to ensure it is completely informed on power infrastructure projects. During these open days, Terna answers citizens' questions, illustrating the reasons behind the investment, as well as presenting project and environmental assessments. This also the occasion to gather opinions and suggestions. In short, a participatory design model is implemented. Every assembly must feature transparent communication, and all aspects of involvement must be underpinned by an agreement that defines exactly what is negotiable and what is not. The goal is to respond to all requests and take care that dialogue is established with citizens and individuals, rather than with organisations. In this context, a participatory design of infrastructure across the territory is the natural path for avoiding and resolving potential conflicts.

2.5 Terna and the Environment

Construction, maintenance or even the simple presence of electricity infrastructure have impacts for the hosting environment. Rather than in terms of use of natural resources or emissions of pollutants, the impact is primarily due to the physical presence of lines and electrical substations which interact with the surrounding environment, both natural and anthropized. Generally, land usage, visual-landscape impact, electric and magnetic fields, and possible interference of the lines with biodiversity (particularly relative to birdlife), are aspects connected to the construction and presence of Terna's physical assets. At the same time, greenhouse gas emissions and special waste are aspects connected to operational activity.

Terna's commitment to containing and reducing the environmental impact of the grid and related activities goes beyond the legal limits where this does not compromise the general interests defined by the concession. This is set out by the company's Environmental Policy, which is fully implemented via an Integrated Management System, and aims at reducing greenhouse gas emissions, implementing energy-efficiency works and mitigation actions to safeguard birdlife.

When grid-development requirements impose creation of new infrastructure, environmental sustainability is considered at every step.

Terna's planning employs careful assessments based on digital thematic mapping, primarily from official sources (regional authorities, water-basin authorities, control-agency network), structured in a vast database that is constantly updated.

Identification of routes or locations for new electrical substations *(Design)* represents the most delicate phase, as this influences the environmental impact for the entire development project. Great focus is given to minimising visual impact, which may involve the use of low-impact single-stem pylons or underground cables, where mitigation is not possible through alternative location or appropriate use of morphological features.

During the construction phase, Terna places importance on identifying access areas and routes to work sites, selected in zones of limited natural interest, when this is compatible with technical and project requirements. At the end of works, Terna performs restoration effort to bring back the areas involved to their original state. If the areas involve natural or semi-natural habitats, Terna carries out specific actions based on soil bioengineering techniques, in addition to standard restoration activities, which can involve the reconstruction of habitats suited to certain plant or animal species, with the planting of native vegetation.

The physical removal of existing lines represents one of the most radical actions taken by Terna to reduce environmental impacts, also in terms of land use. Demolition of overhead power lines is a part of rationalisation work. Between 2010 and 2018, a total of 1,089 km of lines were demolished.

2.6 Partnership with Environmental Associations

Since 2008, Terna has created partnerships with major environmental associations aimed at defining and sharing a balance between respect for the environment and biodiversity and grid development, a notable example of cooperation for sustainable development of the transmission grid.

In 2008, Terna signed an agreement with LIPU (Italian League for Bird Protection and Birdlife International partner) for a scientific study, the first of its kind, regarding the interaction between high-voltage lines and birdlife. The only studies previously published looked at the phenomenon of electrocution, i.e. bird fatalities following simultaneous contact of their wings with the two conductors, typical on low- and medium-voltage lines. The new study supported by Terna, completed in 2011, identified that the risk of birds colliding with electricity lines is low in four of the seven areas monitored. Subsequent addition of observational data, also through new experimental approaches (see the info box on radar monitoring in the Strait of Messina), did not identify particular risks.

In 2009, Terna signed a three-year strategic partnership agreement (subsequently renewed) with WWF Italia in order to increase and monitor the level of integration of environmental criteria within the grid-development planning process, and to plan works of natural restoration in priority areas for eco-regional conservation (through the demolition of existing lines in some cases).

Collaboration with environmental associations broadened over time, with new agreements such as the Memorandum of Understanding signed with Legambiente (a leading Italian environmental association) to promote all necessary initiatives for diffusion of a sustainable-energy culture that brings together development of the electricity system and development of RES. Terna and Legambiente have committed to promoting and spreading knowledge about the world of energy and launching joint actions for environmentally sustainable energy transportation, beginning with the reduction of CO_2 emissions. Regarding the location of infrastructure, Terna and Legambiente have committed to joint mitigation and compensation actions aimed at minimising visual and environmental impacts.

A similar Memorandum of Understanding was signed in 2016 with Greenpeace which, like WWF Italia and Legambiente, collaborates on a strategic level in the drafting of the national transmission grid Development Plan, and on a structural level in drawing up the Strategic Environmental Assessment. Greenpeace is also involved in the implementation stages during local consultation to identify the most compatible location decisions.

These ongoing relationships with leading environmental associations represent a real example of reciprocal exchange of expertise. For Terna, this has translated into greater interaction of its infrastructure with host environments. Three examples are described in the following info boxes.

Box 1 Experimental Radar Monitoring of Birdlife in the Strait of Messina

During creation of the Sorgente–Rizziconi electrical connection between Sicily and Calabria, Terna spent three years monitoring the passage of birdlife over the Strait of Messina to assess potential collision risks for birds migrating from Africa to Europe in relation to the new electricity line. For the first time, in addition to traditional land-based ornithological techniques, monitoring was performed using radar equipment, allowing a census that recorded the passage of over 100,000 birds without any collisions. The radars allowed unprecedented data collection, e.g. on the behaviour of birds in conditions of fog or poor visibility and their ability to change flight path to exploit thermals. Considering the amount and quality of results gathered, of great interest not only to biologists, entomologists, ornithologists and meteorologists, but also important for companies managing power lines, wind-power installations, airports and other similar infrastructure, Terna chose to publish all collected data on its website (www.terna.it).

Box 2 SA.PE.I. and Protection of *Posidonia oceanica*

The SA.PE.I. (Sardinia-Italian mainland) undersea cable, a strategic project for Terna to strengthen the national electricity system, was created in accordance with sustainability criteria. These criteria emerged from the route-definition process. With the exception of the cable landing points, it was decided to avoid shipping routes to minimise the risk of accidental damage to cables and monitoring of sensitive marine areas was established, in particular the “Cetacean Sanctuary”, as well as *Posidonia oceanica* and *Cymodocea nodosa* (two kinds of seaweeds) beds, the latter located near to the mainland landing site of Nettuno (province of Rome).

Posidonia oceanica is a fundamental species for the coastal marine ecosystem and has been classified as a priority for conservation of natural habitats because it represents a vast area with numerous vegetable and animal colonies, and a hugely ecologically diverse environment. To minimise possible impacts on Posidonia beds, in these areas the cable was laid on the seabed without digging into the substratum, instead being fixed with special bracket-type anchors positioned by skilled divers. The choice to use these devices prevents oscillation of the cable that may cause openings in the beds, leading to erosion and consequent environmental changes.

Box 3 The "Nests on Pylons" Project

Terna has long been experimenting alternative uses of electricity lines to support biodiversity. One of the most significant is the "Nests on Pylons" project, created in partnership with the Ornis Italica ornithological association. This involves installing nesting boxes for birds of prey on pylons. Several studies have demonstrated that pylons are frequently used by birds of prey, which are attracted by their height. Birds use them as observation points during hunting and as a resting place, safe from predators.

Constant monitoring of the boxes by Ornis Italica researchers, who verify occupation and tag chicks during the breeding season, allowed a considerable quantity of biological and ethological data to be gathered. It has also allowed identification of a positive effect on biodiversity, demonstrated by an increase in certain species populations, particularly *kestrels*, *scops owls* and *European rollers*. The final component of the "Nests on Pylons" initiative is the "Birdcam" project, which involves installation of webcams on artificial nests for online monitoring of life within the nests. Connection to these webcams also allows scientific observation of the birds' behaviour for researchers working remotely. Finally, in 2015, a GIS census (location using geographic coordinates) of installed nests was launched, with 384 being recorded by the end of 2018.

3 The Energy Transition: Future Scenarios and Terna's Role

3.1 Critical Issues and Possible Solutions Identified by Terna

Today, Terna is taking a leading role in sustainable Energy Transition, using its unique innovations, capabilities and technology to the benefit of all stakeholders. In the context of great change, where decarbonisation emerges as a global target, the electricity grid represents a key enabling factor. In this new scenario, Terna has identified certain critical issues that must be faced:

- Reduction of the so-called "reserve margins" to cover peak loads, due to decommissioning of thermoelectric capacity that currently provides a major contribution (passage from overcapacity to increasing deterioration in adequacy conditions).
- Increase in the frequency and significance of grid congestion, due to non-uniform distribution of RES across the national territory (primarily in the South), lack of alignment between their distribution and load distribution, and the progressive loss of conventional generation capacity.
- Increasing steepness of the evening ramp-up of residual load, i.e. electricity demand net of renewable energy, which has been significantly altered in recent years.
- Growing periods of overgeneration from non-programmable renewables that may lead to increasing loss of power generated by these plants, placing the decarbonisation target under risk (Fig. 9).

To face this situation, Terna has identified several possible solutions to meet the challenges of the future energy scenario and to achieve decarbonisation targets by ensuring full integration of non-programmable renewables, whilst guaranteeing system adequacy and security. Specifically, there are five main categories:

- Investments in the national transmission grid and in interconnections with foreign countries, targeted at strengthening grid meshing, reducing congestion,

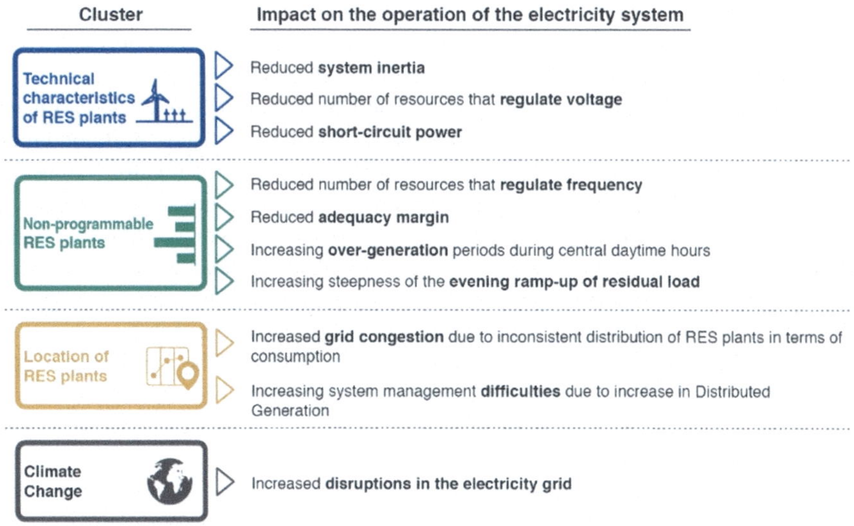

Fig. 9 Critical issues in the new energy context. *Source* Terna

removing restrictions and increasing transport capacity both within the Country and across borders.

- Correction of long-term price indicators in order to build (or convert) new generation systems, both thermoelectric (flexible and efficient gas systems to replace the more obsolete and polluting thermoelectric capacity), via mechanisms such as the capacity market, and renewables through instruments such as Power Purchase Agreements (PPAs).
- Development of further storage capacity. This, if appropriately located, will contribute to guaranteeing the minimisation of overgeneration during the hours of maximum solar generation, providing excellent services to stabilise the grid due to the high flexibility of these systems.
- Integration of the services market at European level and its opening to new resources such as demand response, distributed generation and storage, in order to increase the facilities available to supply the dispatching services needed to guarantee electricity system security.
- Investments in digitalisation and innovation, both in terms of grid management and in terms of real-time monitoring and control of distributed resources.

3.2 The Next Energy Programme

The briefly described Energy Transition not only requires financial resources, but new, qualified and advanced competencies. To support these developments Terna, in partnership with the Cariplo Foundation and Cariplo Factory, has launched Next Energy, a programme that promotes young talent and supports development of teams of innovators, start-ups and innovative companies in areas connected with energy-system development. This initiative, heavily focused on new technologies, reached its fourth edition in 2019, and is composed of three separate calls.

The **Call For Talents** is aimed at new graduates under 28 years old who have earned a degree in Engineering, Mathematics, Physics, Statistics or Economics completed within the previous 12 months, and offers ten paid internships lasting six months at Terna facilities, working on the development of innovative projects.

The **Call for Ideas** is aimed at teams of innovators or young start-ups. Next Energy offers the ten best teams access to a personalised incubation programme for a maximum of two months, after which the jury will award the winning project a €50,000 prize to be spent on acceleration and go-to-market services.

The **Call for Growth** is aimed at more mature start-ups. The top five start-ups will have access to the subsequent "Engage" phase, managed by Open Innovation, Cariplo Factory's platform, with the goal of defining potential areas of synergy with Terna and a pilot project for testing.

All the information on this initiative (e.g. calls, application forms and news) are available on the dedicated website (https://nextenergy.cariplofactory.it/en).

E-Distribuzione: The Role of the Largest Italian Distribution System Operator in the Energy Transition

Livio Gallo and Vincenzo Ranieri

Abstract With a network of over one million kilometres, customers in excess of 30 million, and a workforce above 15,000 people, E-Distribuzione is the largest Italian electricity distribution system operator. After briefly tracing the company origins from the unbundling of formerly vertically-integrated Enel group, the paper offers a detailed overview of the projects the Company has made in the recent past in order to turn its network into a smart grid. This multifaceted upgrade and improvement programme has enabled E-Distribuzione to achieve a more flexible management of the grid, stronger control over the infrastructure—both low- and medium-voltage—, and improved resiliency. Even more importantly, it has been key to the evolution of the network towards an open, decentralised model of power generation and distribution based on ever-increasing reliance on RES. Along this path to Energy Transition, E-Distribuzione has pioneered new technological solutions, conducted intensive R&D activities, both in-house and in cooperation with a variety of university, government, and industrial partners, and launched a number of experimental projects.

Keywords E-Distribuzione · Smart grid · Innovation · Digitalisation · Resiliency

1 History of the Company

E-Distribuzione is the largest Italian electricity distribution company, belonging to Enel Group, that manages the medium and low voltage grids in over 7400 municipalities in Italy according to a concession released by the Ministry of Economic Development (MISE). The company, originally named Enel Distribuzione, was established on May 31st 1999, with the scope of carrying on the distribution and selling activities to captive customers as a spin-off of vertically integrated undertaking Enel S.p.A., pursuant to the legislative decree 79/1999 requiring the liberalization of the electricity market.

A. Gilardoni (ed.), *The Italian Utilities Industry*,
https://doi.org/10.1007/978-3-030-37677-2_13

In 2007, due to the legislative Decree no. 79/2007 that required the unbundling of the distribution of electricity from the selling activity, Enel Distribuzione, through a spin off, transferred the selling activity to captive customers to the newco Enel Servizio Elettrico S.p.A. The Italian Energy Regulator (ARERA) with the Resolution 296/2015 required an enforcement of the functional unbundling obligation of vertically integrated companies, with reference to communication and brand policies. For this reason, from July 2016 Enel Distribuzione S.p.A., changed its name to E-Distribuzione S.p.A.

The electricity distribution perimeter includes the following activities:

- transport of electricity in medium and low voltage networks and its transformation from high voltage network;
- operation, planning, development and maintenance of the distribution network, including connection of customers and producers and metering activities.

The electricity distribution, performed in compliance with the concession regime and the ARERA regulation, is a public and universal service obligation, with the aim of providing equal and non-discriminatory conditions for all customers connected to the distribution network. E-Distribuzione plays an international leading role through the continuous innovation and digitalisation of network infrastructures that guarantee a high-quality distribution service as well as incessant operating excellence.

2 The Network of E-Distribuzione: People, Customers and Infrastructures

E-Distribuzione manages 85% of the Italian electricity distribution network, serving about 31.5 million customers for a total distributed electrical energy of 227.7 TWh in 2018. The company, relying on a workforce of more than 15,000 people, ensures a widespread and extensive presence in the territory through more than 1.15 million km of electrical lines and more than 447,000 of primary (high voltage/medium voltage) and secondary (medium voltage/low voltage) substations.

As of 31 December 2018, over 760,000 producers were connected to E-Distribuzione network for a total amount of about 26 GW of installed capacity from renewable energy sources (RES). Therefore, the company has a crucial role for the effective integration of RES producers, contributing to achieve the national and international strategic targets on decarbonisation of the power system and electrification of consumption. The company represents a global benchmark in the innovative management of digitized electrical grids. As of 31 December 2018, its network is operated through over 200,000 remote controls installed in the medium and low voltage grid that enable near real time interventions from the Operating Centres where a continuous oversight of the infrastructure is carried out.

3 The Path Towards Smart Grids: Main Projects and Achievements

The global trends of the Energy Transition, such as the decarbonisation of power generation and the increasing spread of distributed RES—together with the digitalisation and automation of services and the active participation of communities in the energy market—require the traditional electricity distribution grid to take a step forward towards an open, digital, and resilient smart grid.

As stated by the *European Technology Platform for Smart Grid*, a smart grid is an "electricity network that can intelligently integrate the actions of all users connected to it—generators, consumers and those that do both—in order to efficiently deliver sustainable, economic and secure electricity supplies". To achieve this ambitious goal, a digital and resilient infrastructure, able to cope with the multiple solicitations coming from the social, economic, and political context as well as the environment, turns out to be crucial.

E-Distribuzione is at the forefront of the technological development in the field of smart grid. The company has developed unified smart grid technologies with a "scalable by design" concept and a distributed intelligent architecture, involving remote control, system automation and protection as well as advanced regulation functionalities. With these technologies, it is possible: to maximize the RES hosting capacity, acting on voltage regulation; to clear faults and restore in near real-time the electricity supply; to observe the grid behavior, using a distributed monitoring infrastructure. The research and development activities of smart grid technologies is performed in E-Distribuzione Smart Grid Labs of Milano and Bari, using innovative simulation and testing platforms, such as the "Grid in a Building" real-time digital simulator.

The path towards the state-of-the-art smart grid is the result of a long-term strategy whose pillars are digitalisation of the network, resiliency of the infrastructure against extreme weather events, and sustainability.

3.1 The Digital Network

The digitalisation of the electricity distribution infrastructure has been the fundamental driver of the company development projects. The implementation of smart grid technologies started with the deployment of the smart metering system, which is at the core of the digital business model of E-Distribuzione. The smart metering system represents the first infrastructural layer of the digital network. In 2001, E-Distribuzione launched the Telegestore Project, designing and deploying an innovative proprietary smart metering system and installing more than 36 million remotely-managed smart meters. The architecture of the Telegestore System was based on a two-step data acquisition process: firstly, data are gathered and stored in the smart meters and then transferred using the Power Line Communications

(PLC) to an intermediate device called concentrator. Secondly, the concentrator forwards the gathered data to the Head-End System (HES) of the DSO using mobile communications.

The smart meters were able to remotely read data related to customers' energy flows. Thanks to remote reading, E-Distribuzione has been able to collect metering data for billing and diagnostics information from smart meters in a very timely manner, with no need of on-field interventions. Thanks to the remote data gathering, it has been possible to reduce billing estimations, to introduce time-of-use tariffs, and to ease commercial operations. The smart metering system has also been an important enabler for the liberalization of the energy markets, by providing retailers with the possibility to introduce customised energy tariffs and change them through the Telegestore system. In addition, the new metering system has enhanced the fast activation of the supply service after the signature of the contract. The remote management has enabled the "minimum vital services" (known as "social available power"), whereby customers who do not regularly pay the bill are given a minimum power for essential electricity usages during a limited period of time before proceeding with disconnections. Following the benefits of the Telegestore Project in terms of operational effectiveness and cost efficiency, in 2006 the Italian National Regulation Authority (NRA) made the deployment of smart metering systems mandatory to all Italian DSOs.

The deployment of the first-generation smart meter in Italy has been a pioneering project worldwide as well as the first building block of an industrial strategy relying on technological and process innovation. It led to one of the most advanced automated and remote managed electricity grid, improving the quality of service and fostering the development of Distributed Energy Resources (DER).

By continuing along that road, with an investment of €4 billion, E-Distribuzione launched the second-generation (2G) smart-metering project called "Open Meter", being the first Italian DSO to start the deployment of this new system. Thanks to the excellent performance of first-generation smart meters, E-Distribuzione implemented the 2G smart meters with the same PLC technology, adding a back-up channel, as requested by the Italian Energy Regulator, relying on radio frequency (RF) technology. Moreover, the new architecture includes a new communication channel, known as *chain 2*, that uses a standardized protocol dedicated to the transfer of near real-time data from the meter to the customer by means of in-home devices. The real-time data can be used to improve customer awareness, to support new commercial offers, home automation as well as network ancillary services. *Chain 2* can enable new innovative services like prepaid contracts or demand-response. Therefore, the second-generation smart metering system is one of the key elements of a process of renewal towards a concept of grid where customers can become active participants in the energy market.

Table 1 compares the main parameters of the first and 2G smart metering systems, pursuant to the Resolution no. 87/2016 of the Italian Energy Regulator, containing the minimum functional requirements of the new metering system and a related set of performance indicators. It clearly emerges that 2G metering data have to be collected with a finer sampling and a faster remote reading.

Table 1 Main requirements of first and second generation smart metering systems

Metering data	2G smart metering	1G smart metering
Active energy withdrawn	1 value every 15 min	3 values per month (according to time bands)
Active energy injected	1 value every 15 min	3 value per month (according to time bands)
Reactive energy withdrawn	1 value every 15 min	3 values per month (above 10 kW)
Reactive energy injected	1 value every 15 min	3 values per month
Active power withdrawn	Peak value integrated in 15 min	Peak value integrated in 15 min (peak)
Instantaneous active power withdrawn	On chain 2 up to 1 s	No
Active power injected	15 min (avg)	No
Outages	On event occurrence	Not used due to memory restriction

Fig. 1 Installation of 2G smart meter. *Source* E-Distribuzione

As of October 2019, over 12 million open meters were already installed, while the massive rollout plan will be completed by 2024 (Fig. 1).

The smart meter is only one dowel towards the digitalisation of network infrastructure. The path to digitalise medium and low voltage systems continued with the introduction of remote control as well as grid automation. While high voltage/medium voltage substation remote control started in the 1970s, the first generation of remote control and automation for medium voltage grids started in the same period as the Telegestore Project. In 2010, within an Innovation Programme of the Italian Ministry of Economic Development, in the frame of a project aimed at upgrading the medium voltage network (*Programma Operativo Interregionale 2007–2013*), several devices, such as new sensors, circuit breakers and fault detectors have been developed, tested and installed. This experimentation led to conceiving a new approach to medium voltage grid operations in order to improve the continuity of supply, by automatically identifying faulted branches and re-supplying the healthy part of the grid, thus reducing the impact on customers thanks to remote management.

By continuing along that road, the grid automation and remote management have been extended also to low voltage grid. In 2012, with the project *Res Novae*, E-Distribuzione contributed, in the municipalities of Bari and Cosenza, to the research, modelling and experimentation, on a demonstrative scale, of an advanced management system of energy flows at municipal level, based on the integration of technologies in the Energy and IT sectors.

Thanks also to these initiatives that brought to an extensive grid digitalisation, as of December 2018 all E-Distribuzione primary substations were fully remotely controllable from operational centers, as well as 180,000 secondary substations and about 25,000 low voltage nodes. The grid digitalisation runs in parallel with the strengthening of workforce management integrated tools (augmented reality, smart glasses and asset geolocation). One of the latest application in this field is the resort to thermal-cameras for smart-phones that make it possible to carry out a thermographic inspection with the aim to detect the presence of a hot spot on the electrical component that can be related to thermal stress or obsolescence of the component.

In order to drive and get the benefits of the digital transformation, by exploiting the potential of Artificial Intelligence, Machine Learning and Edge Computing, in 2017 E-Distribuzione launched the DigI&N Italy programme. The main objectives of this project were to identify the possibility of applying new technologies and redesigning business-as-usual processes. The programme has identified about 30 end-to-end macro-processes related to the management of the electricity distribution grid and clustered in three areas: work and investments, operation and maintenance, network commercial operation. According to their level of digitalisation and complexity, processes have been redesigned following three lines of action: a merely process optimization, a zero-based design method, or a *designed by technology* approach.

With regard to the operation and maintenance cluster, one such initiative includes the use of digital image recognition algorithms to analyse photos of electrical components and identify any critical points, thus making it possible to remotely monitor the distribution network through, for example, the use of drones (Fig. 2).

This technology is nowadays the basis of predictive maintenance. In the same direction, E-Distribuzione is testing the complete environmental and electrical sensorisation of secondary substations. The acquired information enables the various platforms to evaluate the network's operational performance, capture weak signals that could lead to potential failures, and send real-time reports to personnel who can intervene promptly, paving the way to the so-called "network digital twin" model.

The network digital twin is an up-to-date and accurate virtual replica of the physical network, its components and their dynamics. It relies on pervasive digitalisation of DSO processes as well as on new technologies that digitise physical assets and make their digital twins accessible and useful. The applications of the

Fig. 2 Photo of a portion of electricity distribution grid taken from a drone. *Source* E-Distribuzione

network digital twin span from operation and maintenance to network design and development, workforce management and training, cooperation with suppliers and technology partners, and interaction with customers and other stakeholders.

The network digital twin is made of four overlapping layers:

- The Foundation layer made of 3D modelling of the grid and its assets;
- The Dynamic device and sensor data layer, with all the real-time information provided by the sensors installed along the grid;
- The Artificial Intelligence layer, where all the data are processed;
- The Human Interface layer where, thanks to virtual-, augmented- and mixed-reality applications, the employees can interact with the digital twin of the network.

Through this representation, it will be possible to work on the network digital twin as if working on the real network, assessing in advance the effect of actions on network components with benefits in terms of both asset management optimization and operating efficiency in infrastructure monitoring.

3.2 The Resilient Network

The recent trends in power systems have definitely changed the way distribution systems have to be planned and designed. A new paradigm is needed in the development of the electrical system, to ensure the best integration of huge amount of RES on all voltage levels and to increase system reliability against extreme weather events. Global climate changes are indeed inducing an increase in both the frequency and intensity of severe meteorological events, even in unexpected locations, potentially resulting in outages in distribution and transmission networks with a relevant impact on customers. In recent years, Italian distribution networks have been largely impacted by:

- heavy snowfall in winter months, often accompanied by strong winds and storm surges along the coasts;
- heat waves in the summer months.

Heavy snowfalls even at low altitudes can lead to ice sleeves with thickness of several centimeters around the conductors of overhead distribution lines, causing mechanical loads to be much higher than the design characteristics required by the actual technical specifications. In winter months, there is also a considerable increase of failures of overhead conductors caused by particularly intense wind gusts, comparable in some cases to force 4 hurricanes. Exemplary in this regard are the events that occurred in the regions of Veneto and Friuli Venezia Giulia during 2018, with over 14 million trees tumbled down by the wind.

During the summer months, heat waves become more and more frequent and intense, characterized by several consecutive days with very high average temperatures and limited temperature range between day and night, preceded by periods of drought. These particular conditions of humidity and temperature prevent the heat dissipation of buried cables, increasing the risk of breakdowns of these types of conductors, especially in urban areas. In this context, the drawing-up of transparent adaptation and mitigation plans, aimed at increasing system security and reliability, has proved a strategic and concrete answer to the phenomenon. Therefore, the Italian Energy Regulator has introduced guidelines for transmission system operators (TSOs) and DSOs in order to improve grid resilience against various meteorological threats. The resilience of the electricity grid is the ability to withstand strong external stresses—such as extreme weather events—containing the effects of these stresses in terms of both the number of users involved and recovery times. The electrical system as a whole must be able to react quickly to the events, limiting the impact of damages and turning back into operation as fast as possible. According to Resolution 31/2018, E-Distribuzione has therefore adopted a specific approach to assess the grid resiliency, applicable to different types of phenomena, which is based on the calculation of:

- the probability of failure of the medium voltage electrical lines in a certain area;
- the damage caused by the event on the supply of electricity, evaluating the number of customers disconnected, taking into account all the possible emergency power supplies available and all the possible remote control operations permitted by the network structure;
- the risk of power failure in a certain area depending on the type of severe events considered.

E-Distribuzione has determined distinct risk indices for the phenomena related to the various possible threats, as different meteorological phenomena require analysis and models with similar but non-comparable methodological approaches. Due to their strong impact on E-Distribuzione network, the phenomena of ice sleeve and heat waves were the basis of the 2018–2020 resilience plan. The action lines implemented by E-Distribuzione can be summed up as follows:

- replacement of thinner bare conductors with more resilient conductors;
- construction of new sections of line to increase network reclosing/meshing;
- implementation of remote-control equipment to quickly minimize the impact of events.

3.3 The Sustainable Network

In order to deal with the uptake of distributed renewable generation, E-Distribuzione has launched several pilot projects aimed at testing new smart grid technologies. The major ones are described hereunder.

In 2011, E-Distribuzione started Isernia Project, the first Italian smart grid pilot funded by the Italian Energy Regulator to test on-field an innovative model for the protection, automation, and management of power generation in the distribution network.

The objective of the Isernia Project was to test, under real operating conditions, new tools in order to manage in a safe and secure way a system with high penetration of distributed energy resources. The project included the installation of a multi-functional energy storage system used for medium voltage ancillary services such as peak shaving, load profiling, voltage regulation, etc.

The project POI MT, funded by the Economic Development Ministry, took place from 2009 to 2013 in Campania, Calabria, Apulia and Sicily, where there was a plenty of RES connected to distribution grids. The scope of the project was to increase energy produced by renewable sources and promote local development opportunities by enhancing energy efficiency (EE). Each region had a well-defined plan to increase the hosting capacity of medium voltage network, giving direct benefit to both distributor and customers.

GRID4EU is a European funded project developed between 2011 and 2016. It is a large-scale demonstration of advanced smart grid solutions with wide replication and scalability potential for Europe, led by six electricity distribution system operators in close partnership with a set of major electricity retailers.

For the Italian demonstrator, E-Distribuzione led the implementation of an advanced control system able to maximize the integration of distributed renewable energy sources in the medium voltage network. Moreover, technologies to manage reactive power-improving voltage control and to reduce energy losses were tested.

EU-SysFlex is an EU Horizon 2020 project started in 2017, that—using the demonstrator developed for GRID4EU, enhanced with the addition of a static compensator directly installed in the primary substation of the demonstration site (Quarto, Emilia Romagna region)—will test the possibility to provide ancillary services to the TSO. The services will be provided taking into account the mutual needs of TSOs and DSOs mainly exploiting DER connected to the DSO grid.

Finally, ComESto (Community Energy Storage) is a project funded by the Italian Ministry for Research (MIUR). It aims to achieve an integrated management of generation from renewable sources and distributed storage in order to facilitate the active and conscious participation of end users in the wholesale and retail energy markets. E-Distribuzione contributed to the project in terms of industrial research, through the development of a tool to perform an automated technical-economic analysis for grid development through innovative algorithms and machine learning.

3.4 The E-Distribuzione Smart Grid

The digital and technological development of the infrastructure, the enhancement of network components against extreme climate events together with the increase in the hosting capacity of RES constitute the fundamental drivers that led to the implementation of E-Distribuzione smart grid. The main projects carried out in the last decade have contributed to the definition of technologies and algorithms that are now implemented. The company has proceeded from conceptualization and experimentation to deployment on large-scale of smart grid technologies.

Started in 2014 and co-funded by the European Commission in the frame of NER 300 Call, the Puglia Active Network (PAN) Project represents a concrete example of the implementation of E-Distribuzione smart grid. The project, whose implementation phase ended in June 2019, concerns the electrical distribution network located in Apulia, a region in the south of Italy, and takes advantage of smart grid projects developed on a smaller scale. Thanks to the PAN Project, the Apulian grid is one of the first smart grids at regional scale worldwide.

The Apulian network mainly consists of a rural environment with a low population density and substantial renewables penetration mainly from photovoltaic distributed generation sources. This massive concentration of renewable energy sources produces strong solicitations on network infrastructure, mainly due to reverse power flows towards the transmission network or possible islanded operation of distributed energy sources, and has required a novel active management of the network.

In the context of the PAN Project, E-Distribuzione has implemented new techniques for the automatic fault detection and isolation, the so-called "smart fault selection" algorithm, with the aim to reduce the number and cumulative duration of long and short interruptions. Furthermore, in order to improve network hosting capacity, an innovative voltage regulation system has been enforced together with the real-time observability of the main parameters of distributed generation. By facilitating sustainable electric mobility through the implementation of electric vehicle charging infrastructure and by fostering customer awareness of electricity consumption through the distribution of ad hoc innovative devices, the PAN Project has put the customer at the center of the smart grid ecosystem.

It is important to highlight that telecommunication technology is the enabling factor of the functionalities implemented. The innovative applications are enabled by a communication infrastructure based on an LTE broadband "always-on" technology that connects medium voltage producers, passive customers, secondary substations and the primary substations in order to realize the "extended substation". Furthermore, communication among all the devices is fully based on IEC 61850, an international standard for substation automation and digitalisation defining communication protocols for intelligent electronic devices at electrical substations.

The main benefits of the PAN Project are related to the possibility to carry out a continuous control of the voltage profile on the medium voltage network, increase the medium voltage network hosting capacity for distributed generation, increase grid flexibility, and optimize power flows, while assuring a high quality of service.

With these features, the Puglia Active Network Project enabled the creation of a sustainable and intelligent energy ecosystem and paved the way for the distribution grid of the future.

On this basis, E-Distribuzione is initiating new projects and activities aimed to facilitate the technological and industrial evolution of the electric system in the context of the Energy Transition, where the DSO is requested to play a key role for the achievement of international and national energy and climate targets.

4 The Future Role of DSOs: From Network Operator to System Catalyser

The role of DSOs is—and will continue to be—the design, maintenance, development, and operation of distribution systems. Nowadays, the transition to a decarbonised energy system as well as technological development, particularly facilitated by digitalisation, is changing this "picture".

The increasing penetration of distributed energy resources and new market players—such as prosumers, aggregators, and active consumers—will usher in a new era. To keep pace with both the transformation of the power sector and the evolution of customer needs, and to take advantage of new opportunities stemming from this new paradigm, DSOs will need to adjust—and expand—their current role.

So far, DSOs have been able to deal with rising volumes of distributed generation leveraging the reinforcement of the grid, but the above-mentioned transformation poses new challenges. Building and upgrading the grid will continue to be a solution—given the need to replace ageing assets and deal with further electrification of heat and transport. Nevertheless, it is a capital-intensive effort with long lead times, thus the current "connect and reinforce" model might not be always the optimal solution for the future.

Further growth of distributed energy resources will lead over time to a less predictable—and often reverse—flow of power in the system, which can affect the traditional planning and operation of distribution and transmission networks. Nevertheless, DSOs have the chance to exploit the distributed flexibility resources connected to their network. By procuring flexibility services such as voltage support and congestion management from their network users, DSOs can optimize both system operations and the need for future grid investments. In this regard, DSOs act as neutral market facilitators, not only by performing their core activities related to the distribution system, but by also providing high-resolution price signals to the market players that own such flexibility assets.

Changing the regulatory framework for DSOs is key for the success of the Energy Transition. A relevant step forward in this regard has been made in the EU, thanks to the "Clean Energy for All Europeans" legislative package entered into force in 2019. According to the new "Common rules for internal market in electricity" directive (namely by Art. 31 "Tasks of distribution system operators," and Art. 32 "Incentives for the use of flexibility in distribution networks"), DSOs will be able to exploit the distributed energy resources for their own voltage regulation and network congestion resolution needs, using local flexibility products/services whose specifications must be defined jointly by distribution system and market operators. Hence, policymakers and regulators should ensure that DSOs are able to oversee, utilize, and coordinate the impacts of flexibility operations on their networks through ad hoc architectures as part of their active system management responsibilities.

In the new scenario, the DSO business model will change: DSOs will continue to serve their local distribution network customers and retain their core responsibilities, but the growing "pressure" to leverage flexibility, and encourage transparent and non-discriminatory market access, will demand new and enhanced functions and activities.

Core functions will remain geared around delivering system security, integrating renewables and ensuring service quality. Nevertheless, against the backdrop of variable renewables integration and the expected penetration of e-mobility, DSOs need greater visibility, advanced monitoring, and control over the electricity flowing across their networks, via smart grids and smart metering infrastructures.

Besides the results achieved, further investment is needed in breakthrough technologies that will accelerate digitisation of the grid. DSOs will need to deepen their understanding and skill sets in technology issues such as data management, systems architecture, robotics, artificial intelligence, and cyber-security.

The DSO model of the future will be grounded on "integration" and "flexibility". Some requirements are immediate—developing a DSO vision statement, identifying the operating model and business design, and establishing skills and investment priorities. Others will be a condition of DSO model maturity over time. In any case, DSOs will play a crucial role in the Energy Transition.

E-Distribuzione is following this approach, contributing step by step to innovating the energy system as a whole, enabling an inclusive, sustainable, and open transition.

Snam: Healing the Climate with Hydrogen

Marco Alverà

Abstract Climate change and air pollution are defining issues for our generation. Current policies to tackle them are not working and CO_2 emissions are rising, which will have severe impacts on our planet. To prevent this, we need deep carbonisation across the world and an approach that transcends the boundaries of nations and energy sectors, whilst also supporting global economic development. Hydrogen could make all that possible. The most abundant element on earth, it can act as a game changer in the energy transition, interconnecting energy systems to create a successful approach. As the analysis advanced in this article will show, Hydrogen is a solution that is definitive, affordable, and global. It is efficient and easy to transport, store, distribute, and use. It can use existing infrastructure. It can help bring renewables into hard-to-abate sectors like industry, heating, and heavy transport. Above all, it can bring more green energy to a growing population, supporting prosperous, productive, and secure lives. It may not be an easy job but the effort is worth it.

Keywords Hydrogen · Climate change · Renewable energy · Energy transition · Green · Gas

1 About Snam

Snam is one of the world's leading energy infrastructure companies. It has been building and managing a sustainable and technologically advanced network to guarantee energy security for about 80 years. Snam operates in Italy and, through subsidiaries, in Albania, Austria, France, Greece, the UK and has started activities in China. Snam is also a major shareholder of TAP (Trans Adriatic Pipeline). It has the largest gas transmission network (over 41,000 km including international subsidiaries) and storage capacity (over 20 billion cubic meters including international subsidiaries) in Europe. Snam also manages the first LNG regasification terminal built in Italy and is a shareholder of two other strategically positioned LNG terminals in the Mediterranean, in Italy and Greece. Through the SnamTec (Tomorrow's energy company) project, Snam invests in innovation and energy transition businesses such as sustainable mobility, energy efficiency and biomethane.

A. Gilardoni (ed.), *The Italian Utilities Industry*,
https://doi.org/10.1007/978-3-030-37677-2_14

Fig. 1 The Hydrogen Challenge: 2019 ESG Conference, 10–11 October 2019, Rome. *Source* Snam

This also includes research and development initiatives for Hydrogen. Snam is a strong believer in Hydrogen and its potential to act as a game changer in the energy transition towards a cleaner world. For this reason, in April 2019, Snam was the first company in Europe to introduce a mix of 5% Hydrogen and natural gas in its transmission network and, in December 2019, increased the blend to 10%. The experiment conducted in Contursi Terme, in the province of Salerno, resulted in successfully fuelling a *pasta* factory and a mineral water bottling company for a month, and generating wide international interest. To further build the momentum around hydrogen, Snam held a major event on the 10th and 11th October 2019 in Rome, opened by Italian Prime Minister Giuseppe Conte, which focused on the prospects for a "Hydrogen Revolution" and the ESG responsibility of companies.[1] Today, Snam continues to advance its commitment to promote the adoption of Hydrogen as a clean energy vector for decarbonisation (Fig. 1).

[1]On the topic, Snam also published, "Generation H: healing the climate with Hydrogen", a book written by the CEO Marco Alverà and published in English by Mondadori, with contributions from international experts such as Gabrielle Walker, Lord Turner, Baroness Worthington, Luigi Crema, McKinsey & Company.

2 The Hydrogen Opportunity

There is a longstanding joke on Hydrogen that says it is the fuel of the future and always will be. Yet, the unprecedented momentum that has been building around Hydrogen today suggests that the future may have come. It is looking more and more likely that this time, the world's most abundant element is here to stay.

With a long history of potential as a clean energy solution, Hydrogen has suffered from false starts and drawbacks, too often sidelined and never succeeding in its adoption as the leading source of energy. But lately, some positive things are happening. The impressive cost reduction of solar and wind power has shown that strong policy support and technology innovation can combine with entrepreneurial drive to scale the development of affordable clean energy. Today a consensus seems to be materialising on the potential role of Hydrogen as a driver of the energy transition.

With climate change mitigation and air pollution being recognized amongst the most pressing issues of our time, people have been mobilizing to fight it, making concrete changes to their lifestyles. Citizens have started to use their votes and investments to ask companies and governments to do better on climate change. From the rise of ethical funds to green finance, we see companies aiming at zero-carbon objectives, and, on a political level, an increasing commitment from Europe to reaching net zero CO_2 emissions by 2050.

In this new global energy landscape, Hydrogen can be a game changer—acting as a connector for the fragmented energy system. Today, the world of electrons and that of molecules is hardly interconnected, but the choice of one technology as the sole solution to the climate crisis is unlikely to do the trick. Hydrogen offers a fresh perspective, acting as a vector that enables many parts of the energy value chain to become interconnected. It has the potential to be an effective, affordable, and global solution, standing alongside renewable electricity and other low-carbon and renewable fuels. It can be a vital source of energy for a growing population while containing climate change. It can also reduce air pollution, which is estimated by the World Health Organisation to kill 7 million people a year and is a huge cost to society in terms of healthcare.

Of course, Hydrogen's adoption does not come without obstacles. One of the biggest hurdles has been cost, but that is changing, with the reduction in the cost of renewable power improving prospects for cheap green Hydrogen. This should encourage us to work through all the other challenges in Hydrogen's path—from the supply of infrastructure to safety concerns—so that we can leverage its full potential.

The belief in hydrogen's potential is what drove us at Snam to start experimenting its use. In Contursi Terme, near Naples, we blended and injected 5% Hydrogen mixed with methane in the local gas high pressure transport network for one month, which effectively allowed two local factories to be fuelled by the element. Our aim is to build on this work and, study the potential to transport a growing percentage of Hydrogen in a blend with natural gas, so we can provide a

Fig. 2 Snam experimentation at Contursi Terme, April 2019. *Source* Snam

future physical network to connect producers and markets. We also aim to provide a network for ideas, policies and technological dissemination. If the world needs to develop green gases, who better than a gas infrastructure company can get things moving (Fig. 2).

Meeting the challenges that emerge from the energy transition and sustainability trends will require new thinking and common will. Our aim is to highlight just how important Hydrogen can be for the future of our planet, and to spur policy-makers, businesses and consumers to start working to realise its potential.

3 The Climate Challenge and the Call for Green Gases

Climate change is the existential challenge of our generation. Scientists have been warning of dangerous global warming for decades, and in recent years politicians and the public have begun to grasp the seriousness and urgency of the problem. Average global temperatures have risen by almost 1 °C over the past century. This deceptively small number masks a host of growing hazards with severe impacts worldwide.

Fighting climate change is therefore a global issue. It doesn't matter where CO_2 is emitted, just the overall quantity. And it is a stock, rather than a flow, issue. What really matters is not how much CO_2 we will emit in a given year, but the total amount accumulated over time. Pollution is a separate issue, mainly local and urban, caused by other gases and fine particles. Despite this, today our efforts on climate change are not delivering the desired outcome. At the moment, the world is not on track to reach international climate goals that aim to reduce carbon emissions quickly and significantly enough to prevent a dangerous increase in global temperatures. Last year, the world's energy-related CO_2 emissions rose by 1.7% to a historic high of 33 gigatons.

There has, of course, been some progress. At the 2015 Paris agreements, most of the world's nations signed up to the goal of limiting warming to well below 2 degrees. Meanwhile, the technology to generate renewable electricity has improved more quickly than anyone expected. The ambitious targets set by Europe drove the industrialization and mass production of solar and wind technology. As a result, the cost of solar capacity fell by 75% from 2010 to 2018. However, these have not been enough, and emissions started to rise again in 2017, reaching the highest ever level in 2018. Some European countries are set to blow through their CO_2 targets also due to higher-than-expected coal consumption. If we carry on as we are, we may be on track for catastrophic global warming.

One of the key obstacles to finding a lasting solution is that the proposals that are being advanced today suffer from three unsurmountable defaults. Primarily, the world is trying to cut carbon emissions using the tool of green electricity, but although this has merits, alone it will not be enough to prevent extreme climate change. Even in a scenario in which we advance our strongest efforts to decarbonise, the IRENA Renewable Energy Roadmap reveals that by 2040, the percentage of electricity in global final energy consumption will not exceed 38%. This means that green electricity alone cannot meet our deep decarbonisation needs. It cannot fully decarbonise industry, shipping and aviation; it puts a prohibitively high cost on winter heating; and it leads to a fragmented response, within the energy sector and across geographies, with individual nations each trying to find their own way to reduce emissions.

Secondly, the current path is difficult to reconcile with population growth, energy access, and economic growth. True, the cost of green power has fallen massively. In many cases it is cheaper than grid electricity, as measured by levelised cost of electricity (LCOE). But this does not take into account the investments in transport and storage needed to use renewables properly. These costs rise along with the percentage of intermittent power in the mix. And in some applications, for instance winter heating and transport, full electrification is more expensive than other decarbonisation options.

Lastly, current proposed solutions are also failing because they are not supported by the global approach that is necessary to solve a global problem. The international consensus that made Paris possible was the product of a singular convergence between the US and China. With the world's two biggest emitters committed to working together, other countries followed. However, this united front has, since, now broken down. American support for Paris has waned, and the relationship between China and the US has become more tense. The results of the COP 24 held in Katowice (Poland, December 2018) revealed low morale in the negotiating teams, with no progress. Ultimately, Paris doesn't translate into a global strategy, but rather a collection of national strategies.

But these key issues that have held us back until now, also give us clues to what we should be trying to do. To find a solution, our pathway should be:

1. **Definitive**: we need a plan for how to actually meet the carbon budget.
2. **Affordable**: to get durable support, we need a solution that doesn't cost too much, and that preserves or even creates employment—especially for the poorest nations and parts of society.
3. **Global**: it is no use if Europe reaches zero in 2050 while emissions from developing nations keep climbing. The whole world must be involved. To enable this, we must find a way to trade clean energy, creating a global market.

In this urgent plan to be developed, low-carbon and renewable gases will play a decisive role considering biomethane made from waste, biosyngas, low-carbon gas with carbon capture and storage, and, most importantly, Hydrogen.

4 Hydrogen Role in the Energy Transition

Hydrogen can help meet all of the needs for the energy transition to succeed. Unlike all its alternatives, Hydrogen is non-exclusive. It unites the world of molecules and electrons, weaving together different strands of the energy system. It can distribute power between regions and seasons, serving as a buffer to increase energy-system resilience. Because Hydrogen can be used to export green energy from regions with ample wind and sun, or from natural gas producers with carbon capture and storage (CCS), it could level out green energy prices and so lead to a fairer global economy. Hydrogen can also release industry from its carbon burden, encouraging economic growth and encouraging countries to sign up to a climate solution. And with green Hydrogen the basis for zero-carbon, guilt-free air travel, tourism can flourish too. Hydrogen even improves air quality because it burns so cleanly.

Hydrogen should therefore not be considered a technology, but a technology enabler. Like an internet of energy, Hydrogen can connect all the sectors of the economy and society to trigger competition and innovation across sectors and geographies and make energy more affordable, available and abundant for a growing global population. That's not to say that Hydrogen will make the energy transition easy. Climate objectives involve an overhaul of the global energy system that will require unprecedented mobilisation of resources—and a huge scale-up of all available options, with all the operational, commercial, financial and policy challenges that this entails.

Hydrogen will not be immune to these challenges—and because it is downstream of renewable energy, and requires specific midstream, distribution and consumption solutions, its development will be dependent on what happens elsewhere in the value chain. But if Hydrogen is not a silver bullet, it can certainly contribute to make the transition easier. And that is an objective worth pursuing, advancing research and investing on all the steps of its value chain: production, transport, storage—perfecting them so as to help Hydrogen succeed in its mission.

4.1 Production

The production of Hydrogen can take many forms. There are actually more than 40 ways of making Hydrogen, but most of the element is extracted, in its purest form, via two main mechanisms.

The first is through the process of electrolysis, which involves the use of electricity to split water into Hydrogen and oxygen. In this process, which was invented by two British chemists in 1800, the reaction takes place in a unit called an electrolyser, with two noble-metal-coated electrodes, separated by a conductive substance called electrolyte or a membrane. Three technologies are employed: the first, alkaline electrolysis, is a mature technology; the second, polymer electrolyte membrane (PEM) electrolysis is currently more expensive, but has strong cost reduction potential via industrialization; and the third, the so-called "high temperature" approach using a solid oxide electrolyser cell (SOEC). This can achieve the highest efficiencies out of the three technologies, but it is difficult to build SOEC electrolysers at large scale due to the size limitations for the ceramic membranes. The carbon content of Hydrogen from electrolysis depends on the carbon content of the used electricity and can be very low if powered from renewables, producing the so-called green Hydrogen. Unfortunately, due to the current high costs of electrolysers, electrolysis is currently used in only about 5% of Hydrogen production.

In alternative, Hydrogen can also be extracted from natural gas and other fossil fuels. The most widely used method for this is through reforming, which creates carbon dioxide as a by-product. Steam methane reforming (SMR) and autothermal reforming (ATR) exploit reactions between hydrocarbons (mainly methane) and water vapour at high temperatures, generating Hydrogen and CO_2. Today, almost 95% of the Hydrogen produced globally comes from these processes of reforming and gasification of fossil feedstocks. Their cost depends mostly on the used feedstock and the produced Hydrogen has different carbon footprints, depending on the feedstock and process. These production pathways could be combined with (CCS) for removing carbon dioxide. Alternative reforming processes, for example autothermal reforming (ATR), could prove useful in this context, as they allow for a higher share of carbon capture. Carbon capture and storage would be needed to make this a climate-friendly option, producing what is known as blue Hydrogen.

There are also newer, greener and more innovative Hydrogen production methods, which are also being investigated, from processes such as methane cracking to "power to liquids". However, these still require high R&D investments and are far from commercialisation.

Fig. 3 Map of the European gas network. *Source* Snam

4.2 Transport

One of the key characteristics of Hydrogen, which adds to its appeal, is that, unlike energy produced from renewables, it can be transported. Indeed, Hydrogen can be sent through pipelines, or carried in tanks as a compressed gas or a liquid. As the experiment undertaken by Snam near Salerno revealed, existing gas networks can carry natural gas blended with some Hydrogen (Fig. 3).

4.3 Storage

Unlike electricity, Hydrogen is cheap and easy to store, and it can be stored in two ways: centralised storage for seasonal timescales, and distributed short-term storage. Seasonal storage involves huge amounts of Hydrogen, and the only solution is centralised, underground reservoirs. Former salt mines could provide the volume required. On the other hand, for intra-day timescales, distributed storage could be placed close to locations with high Hydrogen consumption to absorb peak demand. This could be done within gas transmission and distribution lines, or using silos or tanks at nodes of the gas network.

5 Powering the World with Hydrogen

Given these properties, when it comes to use, Hydrogen has a vast potential. Most importantly, it can offer an effective decarbonisation opportunity in the hardest to abate sectors such as specific industries, heating and heavy transport. Besides, Hydrogen is the only viable way to store renewable power over seasons, allowing to overcome the intermittency problem that has been the main obstacle to their widespread use (Fig. 3).

Of course, Hydrogen is not the only low-carbon gas. Biogas and biomethane also stand out. These can be made from agricultural or urban organic waste, or biosyngas—a synthetic gas made from renewable Hydrogen and CO_2. Low-carbon natural gas is made by capturing the CO_2 from natural gas after it has been combusted. These clean gases share many of the same attributes as Hydrogen, being cheap and easy to transport and store, and able to use existing infrastructure. However, none seem well-suited to meet the challenge, especially limited by the need for land for its feedstock. Meanwhile, Hydrogen has the additional benefit of being infinite, which therefore offers a huge potential to cut production costs and work in various sectors.

5.1 Power Generation

Reducing the carbon intensity of power generation is mainly a matter of swapping thermoelectric generation from coal and natural gas for solar and wind power: electrons for electrons. However, these renewables are intermittent. So, the more you rely on these energy sources, the harder it is to ensure that you do not end up short. You need to have more panels and turbines than would be required in optimal conditions, so as to ensure adequate production levels even when conditions are not perfect. You need to transport electricity from further and further away, on the basis that it is always going to be sunny/windy somewhere. And you need to store electricity, for instance through batteries or pumped storage. These are called integration costs, and they increase rapidly as the share of intermittent renewables goes up. Hydrogen solves this: given its easy transport, Hydrogen allows to cut integration costs by shifting huge amounts of energy between places and times, at low cost. It can also be burnt in power stations to lift the doldrums or meet peak winter demand.

Such dispatchable power therefore improves the stability and security of the power system. Security of supply is much more valuable now than even a couple of years ago. An increasing reliance on high-tech data and communications systems makes the economy more vulnerable to even short interruptions. We need to ensure that critical infrastructure is properly protected, and dispatchable energy from Hydrogen could help to do that.

5.2 Hard-to-Abate Sectors

Broadly speaking, though, power is the easiest sector to decarbonise. What is harder is to reach zero carbon emissions in industry, transport, heating, cooling and cooking, which account for well over 60% of our CO_2 emissions. Scaling up the use of green electricity can help only up to a certain point, but real change can only happen if decarbonisation spreads to the harder-to-abate sectors. The good news is that, even here, Hydrogen can help.

1. ***Industry: Hot Hydrogen***

Cement and steel-making, which by themselves account for almost half of industrial emissions, produce CO_2 by burning fossil fuels to supply high-temperature industrial processes (700–1600 °C). As high-temperature heating is practically difficult and costly to electrify, Hydrogen, biomass and CCS are being considered as alternatives. Here, Hydrogen infrastructure already exists where Hydrogen serves as an input to industrial processes and where it is produced as a by-product, for instance in petrochemical clusters. Hydrogen allows gradual decarbonisation. For example, ethylene crackers do not require big process changes and shifts in safety procedures to switch to Hydrogen, making the shift easier than a full overhaul towards direct electrification, which would require now machinery and often investments in transmission and distribution infrastructure.

2. ***Heating: Fixing Seasonality***

The characteristics of heating demand make it challenging, inefficient and expensive to decarbonise through electricity alone. Even in countries that are not extremely cold, such as Italy and the UK, winter peak energy demand is several times the capacity of the electricity network. What's more, seasonal heating demands are inversely proportional to the energy generated by one of the main renewable sources, the sun. The third issue with electrifying heat is that people would need to invest heavily in their homes, because heat pumps—which move heat from one place to another, and are more efficient than simply burning fuel to generate heat—require high levels of insulation and have reduced efficiency in cold climates. The interventions required are invasive, and total cost would typically be in the order of €200–300 per square meter. In addition, changes in household behaviour are notoriously difficult and slow, also because people don't like to spend money upfront even if the returns might be worth it.

Hydrogen may provide a solution to many of electrification's challenges. Its high energy density allows much lower storage costs. You can convert electricity into Hydrogen during summer, store it underground and finally use it in boilers or generators or in district heating networks in the winter.

Using pure H_2 will require retrofitting the gas grid and installing new boilers and cookers in homes. Although research projects are underway—such as the H21 Leeds City Gate Project in the UK, which suggests that this is feasible, it is of course harder. Instead, blending Hydrogen with natural gas is a way of reducing the emissions of any form of gas consumption, including heating, and depending on the percentage blend, may not require any investments in infrastructure or appliances. It may also provide a route to scaling up Hydrogen production without significant investments in infrastructure.

The experiment advanced by Snam in southern Italy has shown that it is possible to blend 5% of Hydrogen with natural gas in existing gas infrastructure and power factories—a test which resulted in the first ever Hydrogen fueled *pasta*. This has implications not only for heating, but also for integrating green Hydrogen into the broader energy grid. As research continues, Snam is now on track to repeat the experiment by increasing the share of Hydrogen to 10%—proving that blended Hydrogen can work.

3. ***Transport***

Mobility is where the whole Hydrogen craze began. Almost any form of transport can be powered using Hydrogen, by combustion of Hydrogen gas or Hydrogen-based fuels, or by using fuel cells, which convert H_2 into electricity that can power an electric motor. Hydrogen has much higher energy density than existing batteries, providing a similar range to vehicles powered by gasoline or diesel. This makes it especially suited to decarbonising heavy transport, shipping and aviation. It could also provide healthy competition for electric cars and other light transport. Hydrogen can connect the supply systems for gas and electricity to meet our climate challenges. It enables the gas and electricity grids to collaborate, as it can be produced through green electricity and also from natural gas, transported in the gas grid (in blended form, or pure with retrofits), burned for electricity, and used in hard-to-electrify sectors. It can therefore enable the two grids to work as an interconnected energy network, able to carry, store, transform and deliver renewable energy in different forms, continuously optimising for cost and supply security.

6 The Hurdles to Overcome

Yet, before taking over, Hydrogen needs to overcome several hurdles, especially the challenges of safety, perception, infrastructure, and cost. Security is a priority for all those dealing with Hydrogen and despite experiments that reveal its safety in certain contexts, the volatility of Hydrogen could become a concern when we think about distributing it directly into people's homes. There is also the problem of public perception. Hydrogen is not an everyday commodity. You can't see or smell it or buy it at the supermarket. And the word is associated with bomb and explosion, and the disaster that befell the German airship Hindenburg in 1937. We need evidence-based information campaigns to inform the public about the applications and implications of Hydrogen technologies. We should focus not only on climate change, air quality and energy security for coming generations, but also on the immediate benefits that Hydrogen can deliver for individuals, companies and communities.

Infrastructure is also another obstacle. One of the common criticisms of Hydrogen is precisely the lack of specific infrastructure and technology for final consumers, for instance dedicated pipelines, filling stations, boilers and cars. And clearly, this is true. The logjam—where manufacturers don't make appliances because there is no supply infrastructure, and infrastructure companies don't build the pipes because there are no appliances—isn't easy to break. But it isn't as though the world has never built infrastructure before. The way to address the Hydrogen infrastructure challenge is three-fold—and not necessarily sequential.

First, we need to continue and extend our studies and trials on blending, to ensure that blends up to say 10% of the mix are compatible with existing infrastructure. Second, we should create initial demand for green Hydrogen in markets that already exist and don't require new infrastructure and appliances. That can get the upstream costs down to a reasonable level without needing to fiddle with the market or change consumer behaviour too much. Third, we need to work on full value-chain solutions, where demand in one area is aggregated into a cluster, and then this scale of demand is used to justify investments in infrastructure and in appliances. We see significant interest from potential long-term buyers of renewable Hydrogen, who are keen to decarbonise at costs that could be comparable to other energy sources, with high supply security and low-price volatility.

The final obstacle to the adoption of Hydrogen is, of course, its cost—which is obviously an important consideration for any new technology. But in this case, cost is a hurdle that may not be as high as we thought. Today green Hydrogen obtained from electrolysis costs about US$5/kg, equivalent to US$125/MWh. Blue Hydrogen, from fossil fuels and CCS, is much cheaper at about US$2.5/kg ($60/MWh); but it is constrained by CCS capacity, which is not being developed in Europe yet. At these levels, Hydrogen is still relatively expensive compared to other decarbonisation options—except new-build nuclear—and certainly more expensive than fossil fuels in most sectors. Clean Hydrogen development may start to speed up when it gets to US$3/kg and reach a tipping point below US$2/kg, where it becomes cost-competitive in an existing market.

There is lots of room to optimise costs along the Hydrogen value chain, especially in three areas: the cost of renewables, which as we know is falling fast; the cost of capital, with access to cheaper capital for Hydrogen investments; and in driving down the cost of electrolysers.

7 Taking Action: Policy Push

To overcome these obstacles, there must be a policy push which will allow us to get to the tipping point. We have concluded that a quick win is to start by blending Hydrogen with natural gas in the current network. This would use existing infrastructure to deliver Hydrogen to existing markets for natural gas. It doesn't require any behavioural changes, or any investments for infrastructure or industry users. If Europe and Japan blended 7% of Hydrogen into their natural gas networks, that would get us to well over 50 GW of installed capacity. This one measure could be enough to get Hydrogen down below US$2/kg, and tip the Hydrogen snowball over the edge.

There are, of course, other policies to be tested. Grey Hydrogen in industry is another market that could be addressed. Green Hydrogen would be well placed to replace this eventually, as electrolyser and renewable costs fall, but it can get a leg-up. If Europe decided to gradually increase the penetration of green in its Hydrogen mix, to say 10% by 2030, that would require an electrolyser capacity of 15 GW to meet the mix. With production capacity rising and costs falling, green Hydrogen could then also be tested in heavy transport, shipping and possibly even passenger cars.

As infrastructure build is such an important challenge in all this, two more important areas to look at are: industrial clusters, which can aggregate demand for industry that uses Hydrogen as a feedstock and to provide heat; and city projects, like the one advanced in Leeds, which are harder to roll out but bring Hydrogen close to the consumer. Requiring industry participants to produce, procure, blend or sell gradually increasing percentages of green Hydrogen can get us beyond the tipping point without creating disruption and without significant upfront costs for the economies involved.

But, who should do the pushing? Of course, in theory, an international agreement would be desired, ideally setting a global carbon price and highlighting the niches where Hydrogen is already competitive so as to gradually increase its penetration in the market. But given unravelling global consensus on climate change, and the reluctance many emerging economies have to commit themselves to costly energy sources, such a deal seems increasingly difficult to attain.

Another, and better option could therefore be for a group of Countries and regions at the forefront of the energy transition to form a "coalition of the willing" and take it upon themselves to create a framework for the first Hydrogen expansion. These countries could agree to put in place the policies and bear the modest cost required from now to 2030, to drive volumes and lower the cost of Hydrogen technology for other sectors and other countries, to the point where Hydrogen could compete on its own in a host of applications. A coalition of the willing would minimize the overall cost of the transition, exploiting market forces to reach deep decarbonisation. It is also simpler to enact than the massive industrial and social re-engineering implied by Green New Deal policies, which are tied and often set back by their social implications.

Ultimately, a Hydrogen Club of Countries would make for a *just* energy transition, putting most of the burden on the wealthiest countries, who are also those who have emitted the lion's share of the CO_2 in the atmosphere, and asking them to bear the costs of the initial policies to drive Hydrogen growth. The result of this bottom-up would be a low-carbon Hydrogen that is both blue and green, and therefore the most realistic. That is because the two routes to clean Hydrogen will cost different amounts in different regions, and will therefore be preferred

Fig. 4 The future green hydrogen economy. *Source* Snam

differently by each actor. And in this system, blue and green Hydrogen will compete, creating a process in which blue Hydrogen can be a trailblazer for green, opening new markets for Hydrogen altogether (for example replacing diesel in trucks) and then being supplanted by green Hydrogen as the cost of electrolysers falls. Ultimately, if green Hydrogen does take off, it will start displacing fossil fuels, making them cheaper, which in turn will lower the cost of blue Hydrogen. The result: a sustainable system in which carbon is no longer needed (Fig. 4).

8 Final Remarks

Back in 1874, the French science fiction novelist Jules Verne professed, "*I believe that water will one day be employed as fuel, that Hydrogen and oxygen which constitute it, used singly or together, will furnish an inexhaustible source of heat and light, of an intensity of which coal is not capable.*"

Finally, the time could be ripe for that prophecy to come true. Through its characteristics, Hydrogen provides a way of turning the power of the sun and the wind into something that behaves like oil and gas—efficient and easy to transport, store, distribute and use. And Hydrogen is also potentially infinite and totally renewable. It can use existing infrastructure. It can help bring renewables into those stubborn sectors like industry, heating and heavy transport, where electricity is hard to use. Above all, it can bring more green energy to a growing population, supporting prosperous, productive and secure lives.

The task will not be easy but transforming our energy system to cut CO_2 emissions to zero isn't an easy task either. The development of green Hydrogen gives us an option to decarbonise that is compatible with new businesses and new jobs—for a greener and fairer world. As hard as it may prove to be, the effort is worth it.

Italgas: Investing in the Future of Gas Distribution

Paolo Gallo

Abstract Italgas is the most important player in Italy in the distribution of gas and the third in Europe: it manages a distribution network that extends for a total of about 71,000 km through which it distributes over 9 billion cm of gas to 7.6 million users. The Group, including its non-consolidated subsidiaries, holds more than 1800 concessions, with an historical presence in the main cities of the country. Italgas' Strategic Plan for the period 2019–2025—with an overall investment forecast of €4.5 billion—has as main focus the innovation and digital transformation of the company through the digitisation of assets and processes and the development of digital skills for all the employees. This digital transformation programme is combined with the creation of the natural gas distribution network in Sardinia, thus reaffirming the role—as in the past—of a Company that develops the Country's infrastructure, new M&A operations, and further organic growth, in terms of network extension and users served.

Keywords Energy · Natural gas · Infrastructure · Innovation technology · Digitisation · Smart meters · Digital factory

1 Who Is Italgas

Italgas is the most important operator in Italy in the distribution of gas (DSO) and the third in Europe: it manages a distribution network that extends for a total of about 71,000 km through which it distributes over 9 billion cm of gas to 7.6 million users. The Group, including its non-consolidated subsidiaries, holds more than 1800 concessions, with an historical presence in the main cities of the country, including Turin, Venice, Florence, Rome and Naples.

It was founded in 1837. With its over 180 years of history, it is unanimously recognized as the Company that brought gas into the homes of Italians, contributing to the economic and social development of the country. Today Italgas is above all a company that looks to the future, with growth and development objectives and an

A. Gilardoni (ed.), *The Italian Utilities Industry*,
https://doi.org/10.1007/978-3-030-37677-2_15

important investment plan for the progressive extension of the service and the adoption of digital technologies that make network management increasingly efficient and sustainable.

2 History in Brief

The history of Italgas has its roots in a time that precedes the unification of Italy. It was founded in 1837 under the name of *Compagnia di Illuminazione a Gaz per la città di Torino*, the first Italian company specialized in the distillation of solid fuels for the production of enlightening gas. A year later, the city of Turin grants the possibility to use the subsoil for free, thus bringing gas lighting to some public places and to the wealthiest families.

In 1851, when the Turin Stock Exchange opened, the Company's share was among the 7 "Private Funds" admitted to the listing. In 1863, right after the unification of Italy, the company changed its name to *Società Italiana per il Gaz*. At the end of the nineteenth century, the distribution network on the national territory experienced continuous expansion and the company invested in real estate, technological development and the acquisition of several companies in the gas sector, obtaining a solid position in the market on the eve of the first war. In 1900 Italgas was listed on the Milan Stock Exchange and in the 1920s it started a programme to acquire the majority of gas companies operating in some Italian cities, including Venice (1924), Rome (1929) and Florence (1929), with the aim of creating a large industrial group.

In the post-war years, the Country's methanisation began, with the creation of the first urban network for the distribution of natural gas in Lodi, near the natural gas fields discovered in the Po Valley. In 1967 Italgas became part of the Eni group and took on a leading role in the continuation of the methanisation work. From the 1970s on, with the progressive affirmation of natural gas and the development of the import pipeline network, new city networks were built, and the existing ones were modernised.

In 2003, as part of a reorganisation process, Eni decided to delist Italgas shares from the stock exchange and in 2009 Snam took over the entire shareholding. A new group—composed of Italgas, Snam Rete Gas, Stogit and GNL Italia—was created, active in the entire chain of regulated activities in the gas sector in Italy, from transport to storage, from urban distribution to regasification. More recently, and after 13 years of absence from the Stock Exchange, in November 2016, Italgas returned to FTSE MIB of the Italian Stock Exchange, in coincidence with the separation from Snam: an operation carried out to allow the company to make the most of development opportunities related to new tenders for the concession of urban gas distribution service.

Today, after celebrating 180 years of history in September 2017, Italgas continues the process of consolidation of the sector also through the acquisition of small operators, further strengthening its position as the main player in Italy and third in Europe. A process of acquisitions that also saw the entry into the Group of Seaside, one of the largest Italian Energy Service Companies (ESCo), allowing Italgas to become a leading player also in the field of EE.

3 Investment Plan at 4.5 Billion Euros

The Italgas Group's Strategic Plan for the period 2019–2025 has as its main focus the innovation and digital transformation of the company through the digitisation of assets and processes and the development of digital skills for all employees. This digital transformation programme is combined with the creation of the natural gas distribution network in Sardinia, thus reaffirming the role—as in the past—of a company that develops the country's infrastructure, new M&A operations and further organic growth, in terms of network extension and RdP served.

With an overall investment forecast—excluding sector tenders—of €4.5 billion, the plan lead to a significant increase in the capital deployment, equal to €500 million (+12.5%) compared to the previous one, to take the company into the digital age, strengthening its leadership at European level, taking advantage of all development opportunities, and creating value for investors.

Specifically, of the €4.5 billion planned, about €1 billion will be used for technological innovation and digitisation, including the completion by 2020 of the installation plan for the latest generation meters (smart meters), ahead of the deadlines set by the Regulator. More than half (€540 million) of the amount allocated to innovation is directed towards the digitisation of assets and processes. In addition, the massive installation of sensors and remote-controlled valves is planned for the detection of network operating parameters and their remote management. The Chapter also includes the creation of the new Integrated Center for Supervision and Control of the network in real time, whose project was developed entirely in the Digital Factory of Italgas (Fig. 1).

The Digital Factory is the real engine of innovation, the accelerator in the adoption of digital technologies opened at the company's headquarters in Milan in November 2018. About €70 million are foreseen for the new activities of the Digital Factory and for the development of new applications of Machine Learning, Data Analysis, and Augmented Reality.

Approximately €500 million (+11% compared to the previous Plan) are allocated to the continuation of the plan for the construction of the natural gas distribution network in Sardinia, where the Group has assumed the role of main operator acquiring the management of about one third of the basins of users in which the island is divided and the management of the service in all provincial capitals. The investments are mainly destined to the construction of more than 60 new distribution networks in 10 basins and to activities functional to the supply of the networks. To date, a total of about 400 km of pipelines have been built (equal to about 40% of the total of 1100 planned), which also favours the creation of more

Fig. 1 Digital Factory Italgas, engine of the digital transformation of the Group, in operation since November 2018. *Source* Italgas

than 600 new jobs in the induced sector, and about 20,000 pre-contract requests have already been received, demonstrating that Sardinian families are waiting for methane.

€350 million will be used for growth through M&A operations and related technical investments. It should be noted that the targets set in the previous Plan were reached with the acquisition of a network corresponding to around 5000 km and 190 thousand redelivery points for an enterprise value of around €350 million and a total RAB of around €315 million.

A further €350 million euros are planned for investments by Toscana Energia, a company that at the beginning of October 2019 was consolidated in the Italgas Group, following the acquisition of a further 1.98% stake in the share capital from the municipalities of Bientina, Buti, Calcinaia, Casciana Terme Lari and Palaia. As a result of the operation, the share capital held by Italgas rose from 48.68 to 50.66%. Toscana Energia is a leader in the distribution of natural gas in Tuscany, with a RAB at 31.12.2018 of about €833 million. It holds 102 gas concessions, manages above 8000 km of network with 800,000 users and in the last year has distributed over 1.1 billion cm of natural gas. Finally, €2.1 billion are allocated to network development and improvement activities, including the replacement of cast iron pipelines and the modernisation of fully amortised networks. Therefore, the implementation of the planned organic investments will increase the consolidated RAB at an average annual rate (CAGR) of about 4.7% compared to €6.4 billion at the end of 2018, to reach just under €9 billion at the end of the Plan, without considering the contribution of the sector tenders.

A further opportunity for business development is represented by the sector tenders, with dedicated investments estimated at about 1.9 billion euros. The objective is to increase the market share in Italy from the current 34%, including the subsidiaries, to a share close to 45% in terms of active redelivery points, thus going from 7.6 million at the end of 2018 to about 9.7 million in 2025. In particular, approximately €1.3 billion of investments will be devoted to the acquisition of third-party networks in the areas in which Italgas will be awarded, while a further €0.6 billion (until 2025) will be dedicated to the consequent investments in the networks. The positive completion of the participation in the sector tenders and the realisation of the related technical investments will increase the consolidated RAB at an average annual rate (CAGR) of about 7.1% over the Plan period, to reach well over €10 billion by 2025.

Energy efficiency (EE) is one of the drivers on which Italgas also aims in the 2019–2025 Plan. The acquisition of Seaside in 2018 puts the Group in the best position to develop projects aimed at optimising its overall energy expenditure. The new Plan also assigns Seaside the creation of a platform for the management of innovative projects and the possibility of developing appropriate partnerships for the contribution of innovative products and services.

Constant attention is also dedicated to optimising operating costs, with the aim of exceeding the efficiency targets set by the Regulator. In this sense, significant contributions are expected from the company's digital transformation project. In particular, it is estimated that the benefits will amount to approximately 160 million euros over the entire duration of the Plan in terms of reduced operating costs, efficiency on investments, and higher revenues.

The continuous focus on the optimisation of the financial structure will allow Italgas to fully cover, thanks to the significant operating cash flow, both the investment plan and the return on risk capital, also providing the necessary financial flexibility to support development. The ratio of net debt to RAB (considering the associates companies) at the end of 2018 was just under 60% and will remain at this level for the entire period of the Plan. Moreover, the expected robust cash generation will allow rating indicators to be maintained at levels consistent with a solid investment grade area.

By 2025, thanks to both the organic development and the completion of the tenders, revenues are expected to be around €1.8 billion with an estimated Ebitda margin of around 75% while leverage is expected to be less than 60%.

4 Focus on Digitisation, Smart Meters and Network Monitoring

With reference to the €1 billion investment scheduled in the plan for technological innovation and digitisation, the Italgas programme provides for the installation of digital devices/IoT for the massive acquisition of physical parameters of the network and their interpretation through specific algorithms. This will allow real-time monitoring of network operation, alarm management, big-data analysis and predictive maintenance, dispersion research using the most advanced technologies, as well as further process optimization. After the experimental phase, which in 2018 saw the digitisation of 50 pilot networks throughout the country, in 2019 the project was extended to all networks managed by the Group, with the progressive result of further raising the already high quality standards of security of service in favour of the national gas distribution system.

At the same time, as mentioned, the installation plan for smart meters, the intelligent meters, which, as such, represent one of the enabling factors for the wider project of digitisation of the network, is also proceeding. In 2020, the plan to replace all the 7.6 million traditional gas meters managed by Italgas with the same number of latest generation meters will be completed. To date, more 70% of the total has been replaced. These devices have the remote reading of consumption as one of their main functions. In practical terms, this translates into an advantage for final consumers who see their bill increasingly close to real consumption, as well as the ability to adapt their lifestyles on the basis of known consumption in real time.

In this context, Italgas has also allocated new physical spaces to technological innovation with the inauguration, in autumn 2018, of its own Digital Factory in which multifunctional teams develop new IT solutions in Agile and Time-Boxed mode aimed at transforming business processes. This workplace involves Italgas staff and also offers space and opportunities to the ecosystem of small and medium-sized companies that want to develop, with the support of the company, innovative technologies and tools, according to the model of open innovation. Today, the various Digital Factory rooms are working on projects to make operations more efficient, to further improve customer experience, and to implement innovative solutions, including the development of new Machine Learning and Augmented Reality applications to support field activities.

In the wake of the overall and general innovation of all processes, there is also the agreement signed by Italgas and Picarro, a leading U.S. company in the supply of intelligent software for detecting the presence of gas in the air, water and land, for the monitoring of networks. An agreement that provides for Italgas to use Picarro Surveyor in Europe, currently the most advanced technology in the field of early leakage detection of gas networks. The system, named CRDS (Cavity Ring-Down Spectroscopy), consists in a sophisticated sensing technology developed by Stanford University and licensed to Picarro, which has further implemented it by integrating the function of geolocation, machine learning systems, and sophisticated algorithms to interpret a number of environmental factors, including the wind direction or the presence of any isotopes (Fig. 2).

In 2018 the first applications of the new technology involved the networks of the three largest cities served by Italgas, Turin, Rome and Naples, where the total kilometres of network controlled were above 150% the minimum standards required by the regulatory authority (more than 38,000 towards the approximately 15,000 km of the standard). The presence of vehicle search devices and software with a detection sensitivity of ten orders of magnitude higher than traditional

Fig. 2 A Jeep equipped with the Picarro revolutionary technology for the monitoring of networks. *Source* Italgas

systems—parts per billion versus parts per million—allows through CRDS to identify the origin of natural gas leaks even at a great distance. Among other advantages, the new technology does not require the vehicle to follow the route of the underground pipeline (thus solving the problem of possible obstacles on the route, such as parked cars), is able to detect not only the methane but also the ethane, and exponentially expands the volume of the above-ground area monitored (150–200 m wide and 5–8 m high against 1–2 m and 10–20 cm).

In 2019, with the introduction of the network monitoring system using Picarro's technology, Italgas extended the application of this new system to different and very peculiar urban contexts. Starting with Venice, where the natural gas distribution network is incomparable in the world due to the presence of submarine pipelines, along bridges or under pedestrian crossings. The monitoring of the network in the Lagoon is carried out with the help of a boat equipped in a similar way to cars. Even in the case of searching through the channels, the boat can proceed at normal speed, it is not necessary for it to follow the route of the pipeline and also has two GPS satellites that allow to effectively detect the position of the vehicle.

The digitisation of distribution networks in all their components—meters, workforce and staff—in addition to raising the quality of service and allowing greater cost efficiency, is also the precondition for an increasingly deep integration between the electricity and natural gas sectors, through technologies such as Power-to-Gas (P2G) and microgeneration (MicroCHP), whose purpose is to ensure the lowest environmental impact at the lowest cost to the end user.

P2G technology allows the transformation of surplus renewable electricity from wind and solar power plants into hydrogen or synthetic methane gas through an electrolysis process that splits the water molecule into hydrogen and oxygen. The hydrogen produced in this way can be converted into synthetic methane by various methods, including the capture of CO_2 naturally present in the air. The strong point of the technology is the ability to transform the electrical excess that cannot be stored and transported—except with high costs and high pressure losses—practically without limits of time and space, exploiting a network of existing and widespread infrastructures. At the moment, the necessity of converting hydrogen into synthetic methane depends mainly on the fact that natural gas transport and distribution infrastructures, as well as industrial and domestic equipment connected to the gas network, require more or less significant adaptations from a technical point of view for operation with hydrogen. This is due to the different chemical properties of the two elements, which differ in particular in terms of density (hydrogen less than methane) and calorific value (hydrogen greater than methane). The conversion of hydrogen into synthetic methane makes it possible to avoid these adaptations and the related costs and to make the gas immediately available in the network. A concrete alternative that can be used during the start-up phase of the P2G technology, and for a long transitional period, consists in the injection of hydrogen as it is into the network, because mixing it with natural gas or biomethane in percentages between 5 and 20% does not require particular plant adaptations and therefore additional costs.

The MicroCHP, on the other hand, allows the production of energy and calories to be adapted in real time to the requirements of the final user, even at the condominium level, using the most convenient and environmentally sustainable energy mix.

In both cases, the precondition is a fully digitised infrastructure system, able to interpret the user's requests, know the availability, cost and environmental footprint of energy carriers, and propose the ideal combination of electricity, gas and heat to the user.

5 Sustainable Development

In a scenario where the main long-term risk is the worsening of the planet's environmental conditions, the entire energy sector is requested to drive the transition to a more efficient, secure and sustainable system. Natural gas plays a key role in this respect, as a safe, flexible and programmable source, able to power electricity production, support the development of sustainable mobility, and balance the interruptibility of renewables by ensuring greater environmental protection.

With its know-how, Italgas has always been committed to supporting the economic and social development of the communities in which it operates, actively contributing to the creation of increasingly sustainable cities through the constant improvement of service. This is precisely the heart of the company's strategy and its Sustainability Plan, which takes on board the challenges of the Sustainable Development Goals (SDGs) promoted by the United Nations in its Global Agenda for 2030, and sets them out in five strategic pillars with the aim of strengthening its culture of sustainability, putting people at the centre, being recognised by the

Fig. 3 Cars from Italgas operative fleet, the first to be fully converted to natural gas. *Source* Italgas

territory, creating value for customers and the gas market, and contributing to the efficiency and safety of the energy system.

In the last year Italgas has invested in EE projects that have made it possible to implement technologically advanced solutions for the increasingly efficient management of infrastructures, processes and services. The commitment to the decarbonisation of activities has made it possible, in particular, to complete the project for the conversion of the entire company car fleet to natural gas and, among the other goals achieved, there has also been an increase of up to 80% in the share of gas in the energy mix used by the company and, at the same time, a 5% reduction of the carbon footprint produced (Fig. 3).

Great attention is also placed on the issue of safety, which is combined not only in the guarantee of an increasingly efficient service, but also in terms of reducing accidents at work. Thanks to the company's important technological transformation process, which has made it possible to further improve the quality of its service, it has also been possible to record the best accident indices in the history of Italgas, while continuing to pursue the goal of Zero accidents.

In a consistent way, the company has also equipped itself with new tools for comparison and inspiration, starting with its participation in the United Nations Global Compact, the world's largest voluntary initiative on sustainability issues. In addition, Italgas shares have been confirmed for inclusion in several authoritative global stock exchange indexes for the evaluation of corporate social responsibility. In addition, in 2019, Italgas shares were included for the first time in the Dow Jones

Sustainability Index World, an index that globally assesses the ability of the most capitalised companies to measure themselves against the major issues of sustainability.

6 International Partnerships

The important growth that Italgas has recorded in recent years is also measured by the importance of international partnerships that the company has rapidly developed to strengthen its leadership in Italy and Europe.

The first important step in this direction was taken in 2018 with the foundation of the European Association of Natural Gas Distributors—GD4S—created with the intention of representing the main interlocutor of European institutions on issues that impact the natural gas distribution system and, more generally, on issues such as decarbonisation and sustainable mobility. The association with Italgas involves the main European distributors, namely the Romanians of Distrigaz Sud Retele, the Portuguese of Galp, the French of GRDF, the Spanish of Nedgia and the Irish of Gas Networks Ireland.

In 2019, other important agreements were signed at a global level: from the Memorandum of Understanding with the Chinese State Grid Corporation of China, the largest energy utility in the world, to the consulting agreement with the Greek Eda Thess, to support the development of distribution activities of the major Greek operator, to the intensification of collaboration with the French GRDF for the implementation of joint activities and the exchange of knowledge and experience in network management.

In the context of the partnership with State Grid Corporation of China, which includes, among other activities, the promotion in China of the most innovative Italian companies operating in the natural gas distribution sector, a workshop was recently organized in Beijing. The initiative, on the occasion of the fifth China LNG & Gas International Summit, was dedicated to the development scenarios of the energy sector and distribution activities and it was attended by important players in the sector from different countries around the world (China, Singapore, Australia, Italy and Portugal). In Beijing Italgas, in addition to presenting the digital transformation programme that has been affecting the company for several years now and its best practices in the field of sustainable development, was accompanied by its main suppliers who had the opportunity to present themselves and to illustrate their expertise and expertise developed in the field.

7 The Main Gas Challenges

Europe has decided to put at the heart of its economic and environmental policies the objective of full decarbonisation of energy consumption, to be achieved by 2050. This is a challenging objective, considering that fossil fuels account for more than 50% of the energy mix of European consumption, which leads to a transition process designed to involve and transform all the ways in which energy is produced, distributed and consumed.

The main issue on the transition path concerns the effects and possible impacts of new technologies and the digital revolution on the possibility of producing, storing and transporting energy from RES in an increasingly efficient and economic way, while ensuring the continuity of energy supply in a context of widespread production. A flexibility able to follow the fluctuation of demand and guarantee the resilience of the system to possible breakdowns of the information systems that will support the new energy system.

In this context, after a first phase with a strong dogmatic characterisation, in which it was thought that the solution to be aimed at was the total electrification of the system, now it is gaining ground a more realistic vision in which an electrical system based on RES is associated with the gas system. The reasons that support this second approach are basically the enormous costs that would be necessary to adapt the systems of transport and distribution of electricity to a scenario of "full" or "nearly full electrification" and the growing difficulties in reaching the targets of production of RES with the technologies currently available.[1]

For example, at the recent Energy Infrastructure Forum held in Copenhagen in May, 2019, Euroelectric estimated the annual cost of reinforcing and adapting the electricity distribution network to the digital revolution between now and 2050 at €11 billion.[2] On the contrary, in the same occasion, a joint presentation of ENTSO-E and ENTSO-G, the associations that group respectively the electricity and gas carriers at European level, highlighted the increasing possible interactions between the two systems and consequently the economies of scale available, thanks above all to the P2G and "hybrid demand" technologies for gas and electricity, as well as the more traditional G2P technologies.

In this general framework, Italgas position is that in a post-2030 scenario there will be an important space for gas in the European energy supply. Gas, however,

---

[1]Frontier Economics—*The importance of gas infrastructure for Germany's Energy Transition (June 2018)*—http://www.energy-infrastructure-forum.com/doc/[d2-s1-p03]%20DE%20energy%20transition-green-gas-study%20_Frontiere%20Economics.pdf.

BDEW Bundesverband der Energie und Wasserwirtschaft/ZSW Center for Solar Energy and Hydrogen Research Baden-Württemberg (ZSW) *Press Release (26 June 2019)*—https://www.bdew.de/presse/presseinformationen/zahl-der-woche-halbjahres-rekord-erneuerbare-energien-decken-44/.

[2]Energy Infrastructure Forum 2019—Eurelctric Background papers—*Interlinking gas and electricity infrastructure development (24 May 2019)*—https://ec.europa.eu/info/sites/info/files/eurelectric_-_value_of_the_grids_session.pdf.

will tend to increasingly transform itself from a primary energy source to an energy carrier, in competition with other carriers, and its origin will increasingly be from RES. Translating these considerations into numerical forecasts, we expect gas to represent between 30 and 45% of long-term energy needs, with gas fed into the grid becoming increasingly renewable, through significant growth in biomethane and hydrogen produced using P2G technologies, fuelled by surpluses of electricity from RES. It is expected that a substantial part of the gas supply will be composed of hydrogen from MSR (Methane Steam Reforming), a practice very widespread worldwide and that allows to obtain it through the reaction of methane and steam in the presence of catalysts, associated with CCSU technologies (Carbon Capture Sequestration and Usage), or the set of technical devices that allow the reduction of CO_2 emissions into the atmosphere.

For the infrastructures of the gas system, this framework means, on the one hand, being able to count on a long-term perspective of operational continuity, on the other, it poses significant challenges on the technological and strategic level. With regard to the former, it is a matter of making an effort to adapt the network to make it functional and capable of handling extremely variable mixtures of gas—natural, renewable and hydrogen—while maintaining a high degree of reliability and operational efficiency, so as to guarantee the end customer the return of a product that is qualitatively homogeneous also in terms of calorific value. At the same time, it is the infrastructure operator who must commit to the development of the new P2G technologies also through direct investments, so as to encourage the full maturation of the technologies themselves and the consequent entry of market operators.

8 Energy Transition: The Future of Gas and the Circular Economy

For Italgas, the circular economy is expressed through the activation of industrial and infrastructural models able to support the development of the production and distribution of renewable gas produced by:

- Agricultural waste
- Food waste
- Putrescible Municipal Solid Waste (FORSU).

The reference technology for waste treatment is anaerobic digestion in a controlled environment, whose product is a gas with a neutral footprint from the point of view of CO_2 emissions, because what is released in the production and consumption phases has been previously absorbed by the atmosphere and fixed in the biological matrix of the waste.

The strengths of renewable gas consist in the possibility of being distributed to the final user through the existing infrastructure system of natural gas, as it is or mixed in varying percentages with other gaseous products (methane, hydrogen). Moreover, production can adapt to the characteristics of the feedstock available both in urban and rural areas. The limits to be faced in order to allow the full development of the potential of renewable gas are the costs of connection to existing networks, the limited capacity of some peripheral networks to host local production and the existing incentive schemes at national level which are currently limited to the use of biomethane in transport.

These can be addressed and resolved, to begin with, through the adoption of regulatory policies that allow the recognition of network upgrades necessary to absorb the production of renewable gas, including any "reverse flow" from the distribution network to the natural gas transmission network, and the extension of the benefits provided by the decree biomethane also for the introduction into the grid "without a specific purpose of use".

The estimated potential ranges from the 1.1 billion cubic metres per year provided for by the PNIEC, to about 8 billion hypothesised in a MoU signed by Snam, the Italian Biomethane Consortium and Confagricoltura in 2016. It should be noted, however, that other European countries, such as France, expect a potential production of renewable gas by 2050 that can meet the entire need for natural gas and have recently inaugurated the 100th production plant, against only 4 operating in Italy. In Europe, it is estimated that renewable gas (hydrogen and biomethane) can reach an annual production of over 120 billion cm, with savings of about €140 billion.[3]

[3]Source: Gas4Climate Consortium, composed of Snam, Enagás, Fluxys, Gasunie, GRTgaz, Open Grid Europe and TIGF.

Part V
The Role of the Financial System Behind the Development of the Utilities Sector

Sustaining the Energy Transition in Italy: Financing, Policies and the Role of Cassa Depositi e Prestiti

Luca d'Agnese

Abstract Italy has so far achieved good results in promoting the transition of its energy sector to a more sustainable level, building on a base that was already efficient in many respects, such as energy intensity of GDP and specific carbon emissions of electricity generation. As technologies evolve and new, more stringent targets are to be met on a longer time horizon, new models for project development and financing are needed. Cassa Depositi e Prestiti, a long-standing provider of financing to energy infrastructure, identified those needs for change in its Industrial plan and has developed a set of initiatives to support the development of these new models. The initiatives of CDP and other market players, however, need to be sustained by some policy changes designed to establish a clear long-term view for the evolution of regulation. This paper reviews the national strategy for energy currently in place, the emerging challenges to the traditional model for development and financing of infrastructure, how CDP plans to address those challenges and the requirements we envisage for energy policy evolution.

Keywords National targets · Energy transition investments · Public incentives · Offtake agreements · New financing models · Policy development

1 The National Strategy for Energy

a. National targets and changes in markets *(Need for acceleration in energy transition investments, implications of grid parity for investments in RES)*

The global energy market has experienced a relentless growth in consumption over the last thirty years, accelerated by both economic expansion and population growth of the developing countries, leading to an increase of over 70% of the energy consumptions during the same period. Despite the increasing efforts made for the development and adoption of RES and natural gas, the global energy consumption mix remained almost stable between 1990 and 2018. As a result, RES still account for only the 4% of total global consumptions while fossil fuels still cover the vast majority of the energy mix, totaling about 85% in 2018. Due to these major trends,

A. Gilardoni (ed.), *The Italian Utilities Industry*,
https://doi.org/10.1007/978-3-030-37677-2_16

the CO_2 emissions and global temperatures have continued to increase contributing to growing consensus on the need for concrete actions in all the sectors to fight climate change and reduce emissions. The efforts to limit the increase in global temperature and emission levels led to the signing of the Paris Agreement on climate change in 2015, with the objective of keeping average increase in temperatures below 2 °C compared to pre-industrial levels.

The European Union has been at the forefront of the energy transition, and its Member States have undertaken concrete actions to decrease the share of fossil fuels from around 85% in 1990 to about 74% in 2018, while increasing RES to roughly 10%. To fight climate change and emissions in the next decade, the European Union has also developed the "Clean Energy package" to set three challenging targets for 2030:

- Reduction of at least 40% of greenhouse emissions against 1990 levels;
- Share of RES to represent at least 32% of final energy consumptions;
- Improvement in Energy Efficiency, with a reduction of at least 32.5% of primary energy consumption compared to the business-as-usual scenario.

The new president-elect of the European Commission, Ursula Von Der Leyen, has pledged to a Green New Deal with a view to accelerate Europe's goal to become the first climate neutral major economy in the world by 2050.[1]

In compliance with European procedures, Italy has developed its own national plan for the period 2021–2030 (Piano Nazionale Integrato per l'Energia e il Clima 2018, "PNIEC 2018"), renewing and extending its commitment to more stringent targets by 2030 in terms of reduction of CO_2 emissions and higher penetration of RES over gross national energy consumption (revised upwards from 17.4% to 30%, 18% achieved as of today), including a higher quota in terms of electricity generation (from 34 to 55%, 34.5% as of today[2]). In the last ten years, Italy has recorded a continuous contraction of the energy consumption principally driven by the reduction of industrial and construction activities and the improvements reached in the efficiency of end uses.

However, looking ahead new areas for emission reduction must be identified and pursued: energy efficiency in civil and residential buildings, sustainable transportation and deployment of new technologies.

The declared policy targets for 2030 are expected to be achieved via: (i) improved energy efficiency especially in the residential sector; (ii) decarbonisation in the transportation sector via reduction of consumption, penetration of bio-fuels, and electric mobility; (iii) development of new RES generation capacity and refurbishment of obsolete assets in the electricity sector, namely via additional PV (+30GW) and wind (+8GW) plants. Total investments required to meet the

[1]https://www.ft.com/content/7acce5e8-dda3-11e9-b112-9624ec9edc59.

[2]https://www.gse.it/documenti_site/Documenti%20GSE/Rapporti%20statistici/Relazione_annuale_situazione_energetica_nazionale_dati_2018.pdf.

challenging policy targets are estimated to be approximately €180 billion, of which 20% aimed at the increase of installed generation capacity.[3]

The initiatives in the energy transition represent a potential flywheel for the Italian economy. In 2018, the main sector operators have planned more than €96 billion investments in primary energy infrastructures for the period 2018–2030. This investment plan is expected to have a significant impact on the Italian production base with an incremental positive effect on the GDP and a value added on the national economy of €305 billion, considering the whole life of the investment. In addition, the plan is thought to have a positive impact on national occupation in the same period, with the employment of 140,000 Annual Work Unit (AWU) for the construction and operations of the infrastructures.

The development and financing model for some of these investments, however, needs a significant change in some of the segments of the energy value chain, and especially in the area of RES generation capacity.

b. Challenges to the traditional financing model of infrastructure *(Impact of growing market uncertainties, changing profile of sponsors)*

Traditionally, both in conventional and RES sectors, the prevailing financing structures for greenfield assets have been based on long term contractual obligations on the revenue side either in the form of tolling agreements/take-or-pay offtakes or —mainly in the RES sector—public subsidies (feed-in tariffs or feed-in premiums, FiT and FiP). On this ground, investors and financiers could rely on stable, predictable, long-term cash flows on which asset-based financing structures could be built, ensuring high gearing ratios and scope for equity providers to invest in greenfield projects.

Nowadays, the forms of public support to the prevailing RES generation technologies are being gradually diluted (i.e. via the introduction of the following measures: (i) retroactive reduction of incentives in PV sector, (ii) limited capacity allowed to be incentivised, (iii) allocation of incentives based on competitive auctions, (iv) two-way contracts for differences) with a view to their termination (FER 1[4] provides for the allocation of incentives via competitive auctions until 2021 for a limited capacity allotment of 6.23 GW). The rationale for the above is that (i) as the technologies for RES production mature, their capital costs decrease, approaching the so-called grid parity with conventional—unsubsidised—technologies; (ii) the burden of such subsidies is socialised, i.e. passed through to final consumers via electricity price components affecting households' income and companies' competitiveness.

In this framework of decreasing and limited subsidies allocation, the development of greenfield energy infrastructure has slowed down during the last 5/6 years,

---

[3]GSE rapporto delle attività 2018, p. 2.

[4]The legislative decree FER 1 provides the requirements and the mechanisms to obtain public incentives in the energy sector, with the objective of sustaining energy production from RES.

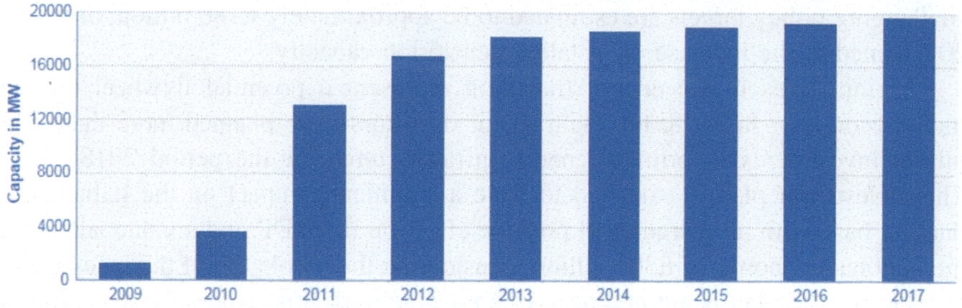

Fig. 1 Italy's solar installed capacity, 2009–2017. *Source* GSE, Inspiratia

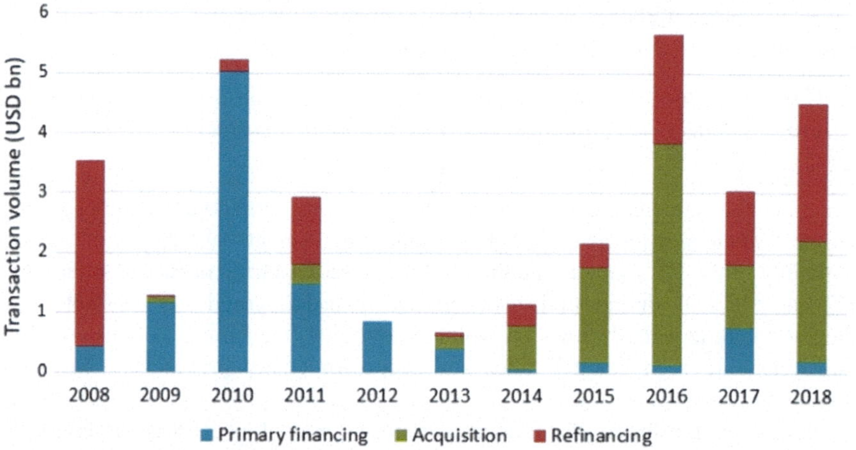

Fig. 2 Italy's closed renewables deals by financing type, 2008–2018. *Source* Inspiratia dataLive

against a surge in secondary market acquisitions and aggregation transactions especially led by financial investors (Figs. 1 and 2).

Given the expected phase-out of public subsidies in the future, the envisaged shift from contracted revenues towards merchant risk brings about uncertainty over cash flows disrupting the paradigm based on which financiers and equity providers have operated so far. Other characteristics of the traditional business model, based on independent developers designing projects making large use of agricultural land, is getting increasingly complex due to the concerns for the use of soil and the impact on landscape of a growing volume of capacity of increasing scale.

In the light of national and supranational targets, there is a strong urge to revive greenfield developments, in order to fill the production gap, via alternative business and financing paradigms.

c. Emerging alternative models *(potential mitigation of merchant risk* via *development of long term PPA)*

A new consensus needs to be built around (i) a new business model and (ii) viable financing structures in the new environment.

With respect to the business model, under the assumption that any form of incentive for mature technologies will be gradually ruled out, market operators are considering a reality where future developments will have to rely on merchant prices and/or offtake agreements (PPAs[5]), if available. Accordingly, the FER 1 decree already envisages future regulation of a platform for negotiations of long-term PPAs, as provided in SEN 2017 and PNIEC 2018.

In Italy, the development of standardised contracts and structures for long term PPAs is still at a very early stage due to (i) the lack of demand for long-term fixed energy prices from corporates, (ii) the scarce demand for such agreements from energy traders. In this context the ability for demand to credibly be bound in the long run to agreed terms and conditions, including the price of energy, is still to be assessed (Table 1).

From a lending perspective, PPAs would entail an assessment essentially focused on the following:

- As regards the creditworthiness of the counterparties, based on conventional financing schemes, the lenders used to deal with revenue flows from Tier 1 Utility or the Gestore Servizi Energetici (GSE),[6] i.e. creditworthy counterparties. With off-take agreements, the lenders would need to assess the credit score of the PPA counterparty for the entire duration of the contract;
- The tenor of the PPA would also play a strategic role as longer tenor stabilises revenue flows in the long term. On the other hand, longer tenor extends over time the reliance on the PPA counterparty (e.g. termination value, etc.);
- The contractual analysis would need to evaluate exit clauses and related indemnities, in order to understand to what extent these would enforce the PPA counterparty to respect its contractual obligations;
- The assessment should also take into consideration the potential alternatives for new off-takers on the market;
- Finally, lenders would focus on the quota of contracted revenues compared with overall project revenues, the higher the quota the higher the reliability of future cash flows.

[5]PPA stands for Power Purchasing Agreement.

[6]Gestore Servizi Elettrici (GSE) operates for the promotion of sustainable development supporting RES through incentives for the energy produced and fed into the grid by RES plants. In addition, it collects, certify and places the energy produced by the incentivised plants on the electricity market. Finally, it evaluates the savings obtained by energy efficiency activities and promotes the production of thermal energy from RES.

Table 1 Private PPA signed in Italy

Company	Offtaker	Date	Capacity	Technology	Status	Duration of PPA (year)
Octopus Energy Investments	Green Trade	June 17	64 MW	Solar	Operational	2
ENGIE Italia	Wienerberger	February 18	n/a	Solar	Operational	5
Octopus Energy Investments	Shell	December 18	70.5 MW	Solar	Under construction	5
Octopus Energy Investments	EGO Group	December 18 (renewal)	64 MW	Solar	Operational	5
Canadian Solar	Trainstone	December 18	17.6 MW	Solar	Under construction	10
European Energy	Axpo	January 19	300 MW	Solar	Under construction/ pipeline	15

Source Inspiratia

2 Role of Cassa Depositi e Prestiti (CDP) in Support of Infrastructure Development *(What Is CDP, Evolution of Its Statutory Role Over Time)*

CDP was established about 170 years ago as an agency for the protection and management of postal savings, investment in projects of public utility and financing of government and public entities. CDP has always played a fundamental institutional role in promoting economic growth in Italy in a sustainable way, pursuing public interest. In 2003, CDP was turned into a private legal entity (joint-stock company) opening its equity to bank foundations that became minority shareholders together with the Ministry of the Economy and Finance ("MEF").

Thereafter, the scope of CDP activity has expanded with the acquisition of a controlling stake in Terna in 2005. Since 2009, CDP has directly financed initiatives of public interest, also in partnership with private entities, and provided support for SMEs, export finance, and social housing. In 2012, the structure of CDP Group was expanded with the entry of several companies, among which Snam reinforcing its support of infrastructure development. In 2014, CDP's remit was extended again to encompass international cooperation, infrastructure financing and investments supporting R&D. Since 2015 CDP has become a National Promotional Institution.

3 Historical Role of CDP for the Energy Sector *(Key Projects Financed, Logic for CDP Presence)*

During the years 2010–2019 YTD (end of September), CDP has supported the energy infrastructure sector with funding up to €6.7 billion (31% of total uses over the period), including both energy and network infrastructures and multi-utilities. In these sectors, CDP has acted via usual financing tools, such as Corporate financing (61%), Guarantees (20%), Bonds (11%) and Project Financing (8%) (see Fig. 3).

Over the years, CDP has been playing a proactive role in the financing of energy infrastructure. The objective of the funding activity has been to provide acceleration in sustainable industrial and infrastructural development. Among the numerous examples and results of the role of CDP for the energy sector, we can mention the funding of the projects belonging to E2i Energie speciali and Piemonte Savoia S.r.l., that can be analyzed in detail also to show the typical structure of projects involving financial investors or multiple industrial players and therefore relying on structured finance.

E2i is a partnership established in 2014 by: (i) F2i Fondi Italiani per le Infrastrutture (managed by F2i SGR, the asset management company founded in 2007 with the contribution of CDP), the largest closed-end fund dedicated to the infrastructure sector with €5 billion in assets under management; (ii) Edison (a leading player in the Italian energy sector with a turnover of €9.2 billion in 2018) and EDF Energies Nouvelles (today EDF Renouvelables, an Italian subsidiary of the EDF Group with expertise in the operation and maintenance (O&M) of RES production plants). In 2014 the partnership resulted in the creation of Italy's third largest operator in the RES sector.

The core business of E2i is focused on the production of energy from RES, mostly from wind power (with more than 700 MW of wind capacity in operation as of October 2019) and, to a marginal extent, photovoltaic (5 MW in operation), with plants mainly located in central and southern Italy. All the plants are directly owned by E2i, without SPVs. As of 2019, nearly 75% of the installed capacity can benefit from price incentives deriving from various forms of incentives granted over the

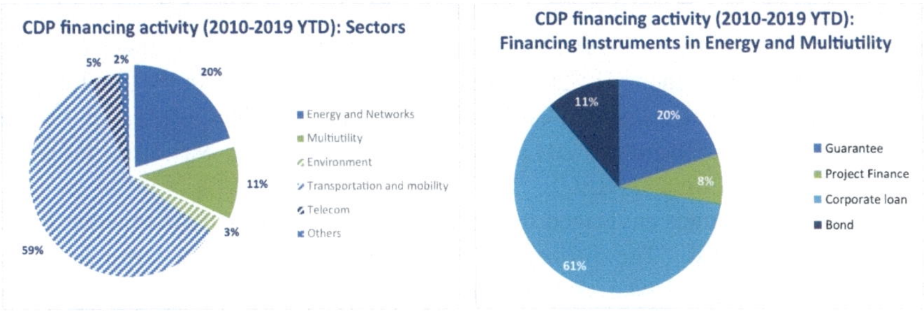

Fig. 3 CDP financial activity in the last decade

past 15 years for the production of energy from RES. The business model adopted by the Company—as concerns both the O&M of existing plants and the development of new plants (greenfield or refurbishment of existing plants)—is based on long-term contracts envisaging the direct involvement of its industrial partners, such as:

- The O&M contract of the existing plants has been assigned to EDF EN Services Italia, with predefined prices until the end of 2024;
- The energy produced by the company's plants is dispatched and sold through Edison according to an off-take agreement in place until the end of 2024. The contract provides that E2i is entitled to receive a fixed remuneration from Edison (computed on the basis of long term P50 producibility scenario) with a protection mechanism to the extent (i) the production variability remains within a predefined collar, and (ii) the availability remains in line with the agreed level. Pursuant to this contract, the risk related to the energy price variability as well as a large part of the volume risk is therefore transferred to Edison;
- Engineering, procurement, construction and management activities for the development of new capacity and the refurbishment of existing plants are regulated by specific EPCM contracts with Edison.

Following the outcome of the auction published in August 2016, the GSE—Gestore Servizi Elettrici assigned new incentivised tariffs to all of the 8 projects presented by E2i for up to 153 MW of additional capacity. The company was able to win the auction which assigned incentives to new capacity of 800 MW out of total capacity auctioned of approximately 2000 MW, thanks to the maximum available discount of 40% offered with respect to the €110/MWh base auction price indicated by the GSE as well as leveraging E2i maximum legal rating, and therefore E2i was entitled to receive incentives with a floor at 66 euro/MWh for its 5 greenfield projects and 90% of such incentives for the 3 full refurbishment projects. With the objective of supporting the investments for the development and construction of the additional capacity (5 + 3 projects, for a total capacity of approximately 165 MW, of which 153 MW incentivised), E2i signed an innovative €100 million unsecured long-term loan with a pool of financial institutions, including CDP. The loan contributed to finance the development and diffusion of state-of-the-art technology in the Italian RES sector, both through installation of new greenfield capacity and through the full refurbishment of existing plants with new capacity, in line with: (i) the European Union targets set in the directive 2009/28/EC on the achievement of 32% quota of energy from RES in the final energy consumption of the European Community; (ii) the principles expressed by the Strategia Energetica Nazionale 2017 (SEN) and (iii) the Piano Nazionale Integrato per l'Energia e il Clima (PNIEC).

The main challenges envisaged in the financing are related to the revenue flows and construction risk. The former is related to potential early termination or non-renewal of the off-take contract. The risk of early termination is properly mitigated by the strategic role of such contract, based on which, among other rights established in the governance, Edison consolidates the company. Furthermore, it is important to underline that as a fallback alternative, the company could: (i) sell its energy on the market with dispatching priority; (ii) continue to receive from GSE the allocated incentives. In this regard, exposure to price and volume risk would be comfortably mitigated by: (i) partial hedging against price volatility entailed in the formula calculation for ex-green certificates; (ii) a feed-in minimum price granted for 20 years as awarded through the auctions to the 8 new plants; (iii) a fixed tariff for 20 years for photovoltaic plants; (iv) past production of the existing plants in line with P50; (v) producibility estimates of the plants certified by an external consultant. As regards the risk of non-renewal of the off-take contract (expiring in 2024), when the incentives will still be due to the majority of the then installed capacity, this is mitigated by: (i) commitment of the company, envisaged in the loan documentation, to discuss a new off-take contract before the expiry of the current one; (ii) acceptable results of the sensitivity analyses assuming merchant revenues with conservative prices after 2024. Other relevant risks involved in the construction of the plants, namely the risks related to delays in construction that may result in the consequent loss of the incentives. These risks were mitigated both by the EPCM contractual structure and by the results of the sensitivity analysis that assumes the loss of 45% of incentives due to potential delays. During 2019, the company has been able to complete the construction works and formally obtain the incentives on all the 8 new plants that have been realised, therefore the risk of losing the incentives is not applicable anymore.

Piemonte Savoia S.r.l. (PiSa) is a project for a transmission facility which received a €342 million long-term loan facility by CDP in pool with other financial institutions. The transaction received the "Europe Power Deal of the Year" award within the Project Finance International 2017 Awards as well as the "Energy Award 2018" from Legal Community. PiSa is a company owned by Interconnector Italia, a consortium company that brings together about 70 industrial entities characterised by intensive energy consumption, including steel mills, cement plants and paper mills. The Interconnector Italy—France is a new electricity interconnection line between Italy and France with a maximum capacity of 1200 MW, divided into two cables with a single capacity of 600 MW each. In the French side of the line, both cables are owned, financed, constructed and managed by RTE, the operator managing the French Transmission grid. In the Italian side, one of the two cables is owned, constructed, financed and managed by Terna (Public Line), while the other cable is owned and financed by the ad hoc vehicle company PiSa, which outsources construction and management to companies belonging to Terna Group (Merchant Line).

PiSa is the owner of the authorisation for the construction and use of the cable for a 10-year period, at the end of which the Merchant Line will re-enter into Terna's ownership according to regulatory provisions. The Interconnector Italy—France was included among the list of Projects of public interest, pursuant to EU Regulation 347/2013 since the realisation of the project aims at supporting the development of the single energy market in Europe.

With the objective of backing the investments for the construction of the Merchant Line, PiSa signed a €342 million secured long-term loan with a pool of financial institutions, including CDP. From a lending perspective, the main challenges that needed to be assessed, addressed and mitigated, are related to construction and market risks. Concerning the construction phase, the main mitigants were identified in contractual provisions such as penalties for the EPC contractor (Terna Interconnector), which guarantee the entry into commercial operations by 31/12/2019, as well as protective mechanisms entailed in the financial structure. With respect to market risk, the project revenues essentially depend on the price differential between the energy price in France and in the northern part of Italy, a factor that could not be controlled by neither of the project counterparties. Although the payment of the debt interest by PiSa is subject to such variability, the risk is mitigated via the protection mechanisms entailed in the financial structures, such as upfront funded reserves and the availability of a junior debt tranche.

4 2019–2021 CDP's Industrial Plan *(New Organisation, Expansion of Scope of Support and New Strategic and Financial Tools)*

In December 2018, CDP approved a 2019–2021 Business Plan that sets out the Group's objectives and strategies in the light of the major economic and social challenges in Italy, the main global trends (innovation and digitisation, Energy Transition and climate change, developing countries and international trade, social changes) and the Sustainable Development Goals of the UN 2030 Agenda. The new Business Plan marks the beginning of a new phase for CDP by strengthening its partnership with the Public Administration through support to Local Authorities in developing and financing projects, also acting as advisor, with the aim of accelerating the development of infrastructures. CDP combines its traditional role of financial provider with that of promoter of new strategic projects, by involving industrial players in public-private partnership operations. The areas of intervention have been expanded, with a focus on mobility and transport, energy and networks, digital, social and environmental aspects.

The increase in activity will be favored by the development of the business model, thanks to CDP's progressively proactive approach aimed at providing effective acceleration, in sustainable terms, to the industrial and infrastructural development in the Country, as well as by the enhancement of the expertise and distinguishing features of CDP: protection of postal savings, long-term investment capacity, complementary role to the banking system and economic and financial balance.

According to the mentioned Plan, CDP will activate a total of €203 billion between 2019 and 2021, contributing significantly to the Country's sustainable growth. It is a significant figure, which will be achieved by using €111 billion of its own resources and by activating €92 billion from private investors and other local and national institutions as well as supranational organisations. All the planned actions will be achieved by ensuring economic and financial balance and, therefore, with full protection of the savings that families entrust to CDP through postal savings bonds and passbook savings accounts.

To effectively support the Country's economic, social and environmental growth, CDP will deploy its actions over four main Lines:

- CDP Enterprises. With roughly €83 billion in new lending over the three-year period and a target of 60,000 enterprises over the horizon of the Plan (reached either directly or indirectly, such as through the banking system) this Line will provide enterprises with dedicated tools to encourage innovation, domestic and international growth and facilitating the access of SMEs to direct and indirect finance;
- CDP Infrastructure, Public Sector and Local development. With a target of investing €25 billion the Line has the objective of supporting local areas and authorities in building infrastructures and improving public utility services, by strengthening its partnership with the Public Administration and opening local offices;
- CDP Cooperation. With €2 billion of target investments this Line has been earmarked for carrying out projects in developing countries and emerging markets. The Plan also highlights a discontinuity in this area envisaging a proactive approach; CDP, as manager of public resources, will assume the role of financing institution, capable of allocating resources through the identification of investment projects;
- Strategic Equity Investments. The Group's portfolio will be reorganised according to an industrial approach and by business sector, in order to support its development over the long term. The objective is threefold: to encourage the creation of industrial expertise in strategic sectors of the production system; to support opportunities for cooperation between investee companies; to support the growth of the different enterprises that come within the value generation chains.

With its new Plan, CDP will proactively contribute to the achievement of the Goals set by the United Nations 2030 Agenda, also signed by Italy. Sustainability will be integrated into CDP's choices through a gradual increase in lending to initiatives whose social and environmental impacts are clear and measurable. According to this approach, new assessment criteria for investments will be adopted for the first time: the criteria that bring together the traditional economic and financial parameters with social and environmental aspects in order to minimise the Environmental Social and Governance (ESG) risk and maximise the positive impacts on communities and local areas. Sustainability, therefore, will no longer be a side effect resulting from CDP investments, which for over 160 years have produced positive aspects for the Country, but a founding pillar in its strategic business choices.

The Plan also envisages the following actions: enhanced cooperation with the Public Administration to relaunch investments and innovation, also through renegotiations and advances to facilitate access to national and European funds and the payment of debts to enterprises; the increase in direct intervention in local areas, with the launch of City Plans (Piani Città) to upgrade urban areas, and initiatives to support tourism (fund for the revamping of tourist facilities, with a special focus on southern Italy), art and culture; the support for public services such as health (healthcare innovation and senior housing), housing (social housing) and education (student housing and student loans) services.

5 New Support Model for Energy Infrastructure *(Combination of Debt and Equity Tools, Initiatives in Support of New Business Models)*

Considering the changing market environment in which allocation of incentives based on competitive auctions is being ruled out, the asset-based debt financing structures used so far will no longer be applicable and, if possible, a new paradigm of financing should be adopted.

A consensus around viable structures has not been found so far, however in a context where full revenue exposure to merchant risk is considered not viable by the providers of debt financing—assuming that at least a portion of such revenues could be contracted under an acceptable long term PPA—potential alternatives might include, by way of example:

(i) Financial structures based on prudential recognition of revenue flows and conservative debt sizing *criteria* leading to lower gearing ratios. The prudential recognition of revenue flows consists in (i) assessing future revenues of the plants until a target date slightly longer than the expiry of the PPA but still shorter than the residual operating life; (ii) prudential assumptions in relation to plants' production levels (iii) prudential assumptions on power price over the period beyond the expiry of PPAs. With reference to debt sizing criteria, the debt capacity will be defined by using higher cover ratios, while potential cash-sweep mechanisms might be in place in case the PPA is not renewed until the target date;

(ii) blending of merchant projects and subsidised assets in order to take advantage of portfolio effects. In this way, companies can leverage a relevant quota of subsidised assets benefitting from stable cash flows in order to slightly increase debt capacity at a portfolio level.

Otherwise, large and diversified corporates might lead the energy transition if willing or able to take on such risk on their balance sheet.

According to this approach, CDP intends to play a proactive role in favouring and accelerating the transition towards more sustainable energy systems. In line with its 2019–2021 Industrial Plan, CDP has been expanding both its operational

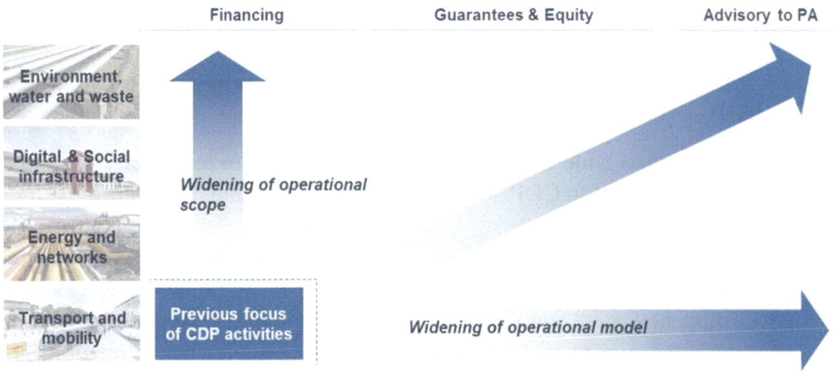

Fig. 4 CDP Public Sector and Infrastructure's scope of support

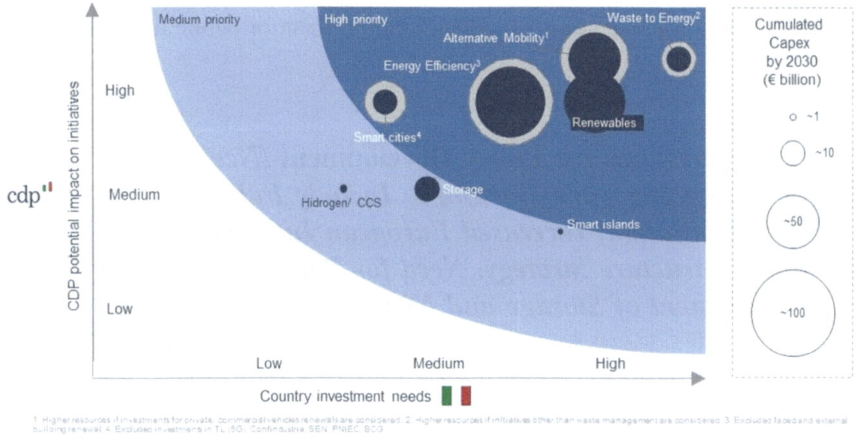

Fig. 5 Prioritisation of the operational scope

scope and its operational model, envisaging also the use of equity, guarantees and advisory services to local entities in addition to its traditional lending operations. Within this new framework CDP intends to pursue, with a focus on sustainability, projects aiming at filling the infrastructural and business model gaps that appear to slow down the deployment of initiatives that will sustain the reach of 2030 targets (Fig. 4).

More specifically, the criteria for identifying the sectors to prioritize, are: (i) the investment gap between the Italian objectives (i.e. PNIEC, 2030) and the initiatives already planned by the sector operators; (ii) the industrial feasibility of the single initiatives (e.g. availability of proprietary technologies); (iii) CDP's value added to the success of the initiatives leveraging its role as national promotion agency and its relationships with local institutions; and finally (iv) CDP's role as long term investor in sustaining innovation and deployment of new technologies (Fig. 5).

For this purpose, in 2019 CDP launched several initiatives with major industrial partners in the sectors in which a consistent effort is still needed in order to reach the challenging energy transition targets set for 2030. The investment areas identified by CDP cover:

- Renewables, with a wide infrastructural gap to be filled to reach the 2030 target and the opportunity for CDP to support investments and leverage underused state-owned land;
- Circular Economy and waste-to-fuel, with a wide infrastructural gap and the opportunity for CDP to sponsor innovative Italian technologies;
- Energy Efficiency, with unexpressed market potential and the opportunity for CDP to play the role of system orchestrator and facilitator in the relationships with the PA;
- Transportation, with unexpressed market potential due to the financial limitations of local transport companies and the opportunity for CDP to modernise local public transport with initiatives aimed at renovating the fleet and supporting infrastructure construction and improvement of public utility services.

6 Open Questions for Policy Development *(Need for Expanding Time Horizon to Include Full Decarbonisation, Increased European Integration of Infrastructure Strategy, Need for Policies to Support Development of Storage and Network Services)*

A clear, stable framework for policy and regulation is a key element to ensure the optimal financing conditions of energy projects. The removal of direct subsidies to production of RES does not make the stability of policy decisions less relevant. Key policy factors affecting demand and economic sustainability include, by way of example: regulation and contracting of storage capacity, policies to stimulate electrification of transport and heating, permitting procedures for new (RES) capacity as well as for extension of thermal power plants, regulation to facilitate exchange of energy of small producers among themselves and with the grid.

It is difficult to anticipate in detail the outcome of all policy decisions that may affect infrastructural investments in energy over their technical and economic lifetime. A reliable guidance on the direction that such decision will take may be provided by the adoption by government authorities of a comprehensive policy framework for the attainment of long-term decarbonisation targets at country level. Italy has adopted a suitable framework with the 2030 horizon, but the lifetime of projects conceived in 2020, and therefore starting operation in the following years, calls for a significantly longer horizon.

Decarbonisation targets set by European Union for 2050 are substantially more aggressive than those of 2030 (−80% of emissions vs. −40% with respect to 1990 base), not achievable through a mere extrapolation of industry decarbonisation trends in the previous years. They require instead an extensive decarbonisation effort in the economic sectors that have so far been impacted in a more limited way, such as transport, climatisation of buildings, energy intensive industrial processes.

At the national level, an extension of PNIEC, or a less binding but equally in-depth effort, to 2050 may be the solution. It will support and orient investment decisions for key infrastructure as well as the development and integration of (relatively) new technologies such as use of hydrogen in transport and industrial processes, new RES leveraging resources with availability patterns different from those of wind and solar capacity, alternative storage technologies.

Another area for new policy development is the need for a greater integration of infrastructure at European level. Increased variability of RES generation affected by weather patterns, massive shut-downs of high carbon thermal plants, changes in traditional gas flows across the continent and integration of new fuels (hydrogen?) in gas pipelines represent areas for the development of new approaches to EU-wide infrastructure integration. Integration of technologies for local and large-scale grids is another challenge for policies regulating power and energy exchanges as well as other common European issues such as cybersecurity of European infrastructure. Policy decisions in such areas will have effects on the need for and sustainability of investments in all the energy value chain.

An additional way for regulation to support investments in energy transition is the development of market for energy services such as capacity (contributing to system adequacy), balancing and backup. Those services, when paid through medium/long term contracts with a central counterpart, may offer a valuable enhancement to the long-term visibility of revenues for new investments, further enhancing the accessibility of efficient financing.

Another complex issue that needs to be addressed at policy level is the way taxation and system-related charges (such as the payment of past subsidies) affect different forms of energy (e.g. electricity and gas) and different generation and consumption structures (such as self-consumption, peer-to-peer networks, etc.). Changes in this area may affect the relative competitiveness of alternative infrastructure and make investment future profitability less predictable.

A broad policy orientation to be adopted may be to remove specific subsidies and asymmetries which do not support decarbonisation. Several exceptions to this rule, however, may continue to be in place. An important contribution to long-term policy formulation may be to describe and make directional choices on areas of exception to cost-reflective, decarbonisation oriented regulation, highlighting areas where asymmetries are deemed important to preserve industrial competitiveness, support vulnerable consumers, and stimulate new technologies.

Financial and Bank Systems Supporting Utilities Industry Growth

Luca Matrone, Matteo Balasso and Urbano Maria Cremona

Abstract The Energy sector is experiencing the most transformational evolution of the last decades, in which renewable energy sources are gaining massive importance and in which local utilities and energy companies are changing their business models and strategies in order to adapt to a new environment. The Italian financial and banking sector has to play a pivotal role in helping the changeover of the energy landscape. Indeed, although there are challenges ahead, the efforts of some financial players like Intesa Sanpaolo in facilitating the development of concepts such as sustainable finance and circular economy, are encouraging indicators of a well-established trend towards the achievement of global environmental targets.

Keywords Mergers and acquisitions · Market consolidation · Sustainable finance · Financial investors · Brownfield projects

1 Intesa Sanpaolo Group in Few Words

The Intesa Sanpaolo Group is one of the leading banking groups in Europe, with a market capitalization of €40.3 billion as of 5th November 2019; it is engaged in supporting the economy in the countries in which it operates, where it is also committed to becoming a point of reference in terms of sustainability and social and cultural responsibility.

The Group is the leader in Italy in all sectors of activity (retail, corporate and wealth management). It offers its services to 11.8 million customers through a network of approximately 3900 branches throughout the country with market shares of not less than 12% in most regions.

Intesa Sanpaolo has a strategic international presence, with approximately 1100 branches and 7.2 million customers, including subsidiary banks operating in commercial banking in 12 countries in Central and Eastern Europe and in the Middle East and North Africa, and an international network specializing in corporate customer support in 25 countries, particularly in the Middle East and North Africa and in those areas where Italian companies are most dynamic, such as the United States, Brazil, Russia, India and China.

A. Gilardoni (ed.), *The Italian Utilities Industry*,
https://doi.org/10.1007/978-3-030-37677-2_17

Intesa Sanpaolo has always considered environmental management to be a fundamental part of a wider management model which embeds CSR across the entire Group. Numerous initiatives are focused on the mitigation and careful management of direct and indirect impacts on the environment. In particular, Intesa Sanpaolo was the first Italian bank to debut as a green issuer with the 2017 €500 million Green Bond. Moreover, the Group shows a strong commitment to the circular economy as a way of reducing the environmental footprints of businesses and individuals. In this matter, in addition to the partnership with the Ellen MacArthur Foundation, the Bank established a credit plafond of €5 billion and launched the first Circular Economy Lab in Italy.

2 Evolution of the Italian Local Utilities: From Stock Market Listings to Industry Consolidation

The local utilities' sector has gone through different phases in the last 20 years. Following the privatizations of the main state-owned enterprises, which began in 1994, the local municipalities started opening their local utilities' share capital in 1996 with the Initial Public Offering (IPO) of the company AMGA, completed by the Genoa municipality. Subsequently, in 1998 and in 1999, Aem and Acea, controlled respectively by the municipalities of Milan and Rome, accessed the market. In 2000, the Turin and Trieste municipalities completed the privatization of, respectively, AEM Torino and Acegas. Later on, in 2003, the IPOs of Hera and Meta were realized.

There were several reasons behind this process. First of all, the opportunity for the local municipalities to raise financial resources, reducing the capital employed in a industry attractive for the private sector. Then there was the aspiration to increase the level of efficiency and transparency, placing these newly public companies under the screening of the financial markets. Moreover, through the IPOs, the companies benefited from additional options to finance their development plans, either through equity or debt. Furthermore, thanks to the process of going public, the local municipalities benefited from the liquidity generated by the sale of the utilities they owned.

Finally, the IPOs were instrumental in involving local residents, to whom some shares tranches were dedicated, in a mid-term development plan with benefits for the local community.

In 2000, a new phase started, which lasted until 2009, that saw many companies merge, creating some regional players such as A2A, Hera, Iren, AcegasAps, and Ascopiave. A new consolidation wave then started in 2013 with the combination of Aew and Sel, and the acquisition of 51% of Linea Group by A2A. In July 2019, a new transaction was announced between Hera and Ascopiave. The operation consists in the exchange between the two companies of assets of equal value, crucial for the development in the energy retail activities, on the one hand, and in the gas distribution, on the other, in line with the strategic lines of the two Groups.

Thanks to the mergers, the local utilities reached a critical mass, obtaining economies of scale and scope, optimizing their operating structures, increasing efficiencies, and strengthening the commercial effort to protect the customers' portfolio.

Intesa Sanpaolo, through Banca IMI, was involved in almost all the IPOs and mergers completed by the local utilities (Figs. 1 and 2). It was a very difficult process, that reshaped the industry structure. There were several obstacles and

Companies	Date	Description	Banca IMI's role
Gruppo HERA	2003	IPO	Co-Global Coordinator
Meta	2003	IPO	Global Coordinator & Sponsor
ASM	2002	IPO	Joint Global Coordinator & Co-Sponsor
AEM	2000	IPO	Global Coordinator & Sponsor
acea	1999	IPO	Joint Global Coordinator & Sponsor
acsm como	1999	IPO	Global Coordinator & Sponsor
AEM spa	1998	IPO	Joint Global Coordinator & Sponsor

Fig. 1 Banca IMI's involvement in the listing of Italian utilities. *Source* Intesa Sanpaolo

Companies	Date	Description	Banca IMI's role
Acsm Agam, ASPEM, Azienda Energetica Valtellina Valchiavenna, Other A2A Assets	2018	Merger of Acel Service, Aspem, Azienda Energetica Valtellina and other assets of A2A in ACSM Agam	Advisor of ACSM Agama
a2a, LGH	2016	Acquisition by A2A of a 51% stake in LGH	Advisor of A2A
aew, SEL	2015	Merger between Azienda Energetica Bolzano (AEW) and SEL	Advisor of AEW
AMGA, Gruppo HERA	2014	Merger between HERA and AMGA Udine	Advisor of E.ON
enìa, IRIDE	2009	Merger between Iride and Enìa	Advisor of Iride
agam, acsm como	2008	Merger between ACSM Como and AGAM Monza	Advisor of ACSM Como
ASM, AEM spa	2007	Merger between ASM Brescia and AEM Milano	Advisor of ASM Brescia
AMGA, AEM	2006	Merger between AEM Torino and AMGA Genova	Advisor of AEM Torino
LGH	2006	Merger between AEM Cremona, ASTEM Lodi, ASM Pavia, COGEME Rovato and SCS Crema	Advisor of AEM Cremona, ASTEM Lodi e SCS Crema
Meta, Gruppo HERA	2005	Merger between HERA and Meta Modena	Advisor of HERA
ACEGAS, APS	2003	Merger between ACEGAS Trieste and APS Padova	Advisor of ACEGAS Trieste

Fig. 2 Banca IMI's involvement in the consolidation process of Italian utilities. *Source* Intesa Sanpaolo

complications to be addressed, not only related to financial elements—such as business plans, valuations, exchange ratios—but also to governance, organization and group structures, transactions' structures, location of the headquarters and of the operative businesses.

After the mergers, those companies took a few years to integrate the different cultures and organizations, to share the best practices of each merged company, to leverage the larger clients' portfolios to cross-sell other services, and to implement integrated commercial plans and customer care activities. The current landscape of the sector is characterized by a few players that have reached a significant financial scale evolving from a regional to a national geographic scope, and a number of mid-sized players that maintain a strong local focus on the original territories (Fig. 3).

Meanwhile, new macrotrends arose. The decarbonization process, the technological evolution, and a new collective awareness about sustainability put the customer at the center. In general, the sector responded to changes in an impressively dynamic and innovative way, thanks to the outstanding quality and the extraordinary visions of the local utilities' management. They fostered a deep change in their companies following new paradigms as Energy Transition, New Downstream and Sustainability in their business models.

Most of the companies invested heavily to reduce the environmental impact of their activities, for example by increasing the recycling of the waste collected, by digitalizing and upgrading the networks, by reducing losses in the water systems, in the development of renewable capacity and in the energy efficiency space. They

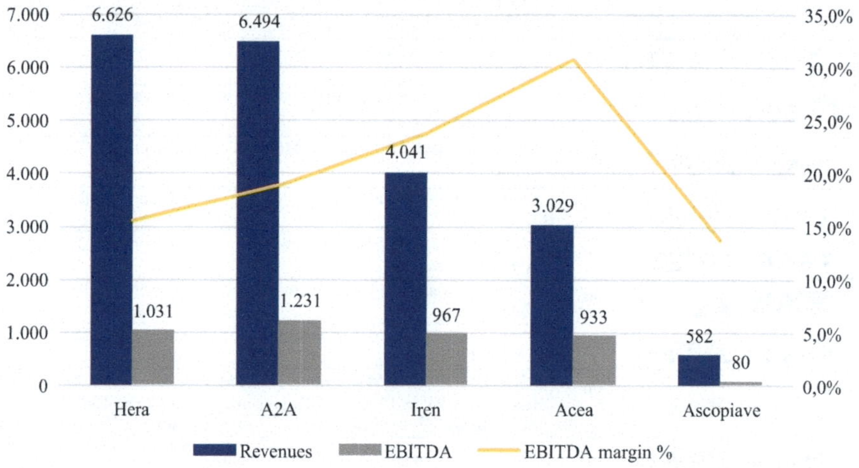

Fig. 3 Major Italian listed local utilities 2018 rankings by revenues and EBITDA (€ million). *Source* Company information

also invested significantly in innovation, in the digitalization of their businesses and in the competence of their workforce, fostering a proactive attitude and a positive culture.

In our view, the time is now ripe to look at more ambitious targets. The champions of this industry should consider new options of aggregation, increasing their size and financial muscles, to be able to face foreign markets, with the objective of expanding, replicating the models they have established and strengthened in Italy.

3 The Birth and Growth of the Renewable Energy Sector

In the past years, the Italian government has brought particular focus on promoting the utilization of renewable energy sources with the objective of increasing their share in the total power generation capacity of the country. This trend has fortified throughout the years thanks to an increased environmental awareness which focuses on reducing carbon emissions, increasing energy efficiency and, more in general, developing a more sustainable energy market and economy.

In 2010, the Italian National Renewable Action Plan (NREAP) had set the target of reaching 17% share of consumption satisfied by renewables energy by 2020. This target was already achieved in 2014, reaching 18.3% in 2017 (Fig. 4).

The investment trend in RES installations can mostly be explained by favorable policies and incentives provided by the government, characterized by a multiplicity of mechanisms that have followed throughout the years in a logic of progressive market orientation and reduction of the incentive level with the decrease of generation costs. In particular, the so-called Conto Energia, which was first issued in 2005 with the approval of the Ministerial Decrees of July 28th 2005 and of

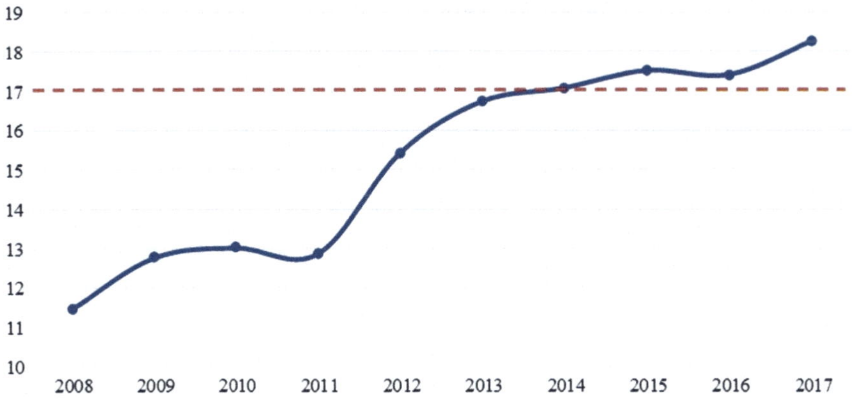

Fig. 4 Share of renewable energy in gross final energy consumption (%). *Source* European Commission

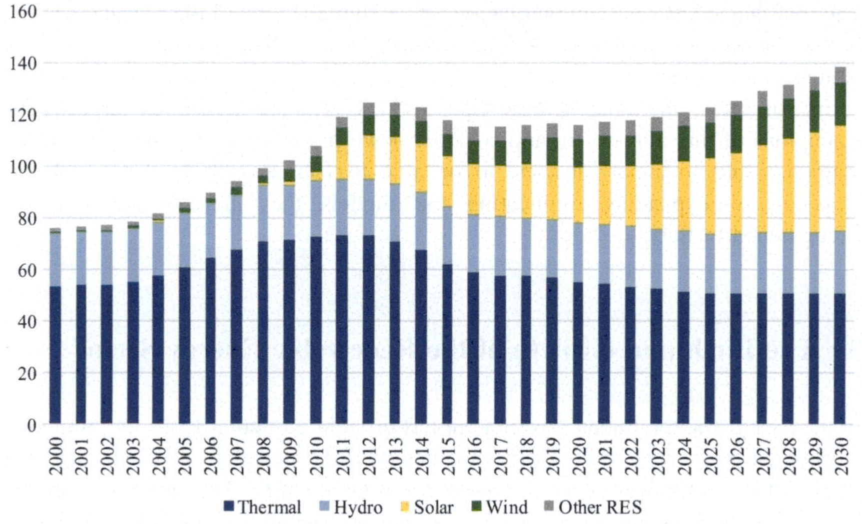

Fig. 5 Development of total installed capacity (GWp) for the period 2000–2030. *Source* GlobalData

February 6th 2006 and later on revised (II Conto Energia on February 19th 2007; III Conto Energia on August 6th 2010; IV Conto Energia on May 5th 2011), was the main driver of the wave of investments in solar PV installations during the 2009–2013 period.

However, the implementation of the incentives' scheme granted to foster the renewables, caused an increase in the electricity expenditure for final consumers, both industrials and households. In order to limit such effects, in June 2014, the government adopted the Law decree nr. 91 2014, amending the incentive scheme to reduce the financial burden for the final consumers.

Italy's renewables market continues to offer opportunities for investors and consumers, also thanks to the new bonus for domestic photovoltaics and the auction system and incentive schemes for renewables, which were approved by the government in July 2019 with the objective of allocating 4.8 GW of renewable capacity for large-scale projects through seven bidding rounds.

Renewable energy sources account for c. 50% of total installed capacity in Italy as of 2018. The RES share has constantly increased y-o-y with a CAGR for the period 2006–2018 of c. 7.2%. On the other hand, conventional thermal energy in the same observation period has decreased with a CAGR of c. −1.0%. As shown in Fig. 5, most of the decrease in thermal capacity can be identified in the 2013–2016 period.

Indeed, until 2010, installed capacity was mainly represented by fossil fuel-fired plants, which represented c. 71% of the total. With the European Commission energy targets and new incentive schemes, installations of new RES increased exponentially. In particular, in the 2009–2013 period, Italy experienced a drastic

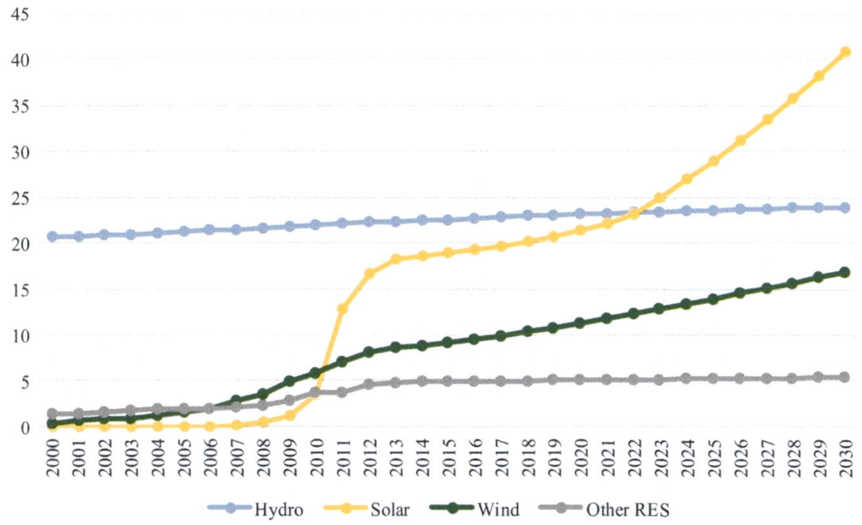

Fig. 6 Development of renewable energy sources installed capacity (GWp) for the period 2000–2030. *Source* GlobalData

increase in solar power plant installations, becoming one of the major solar photovoltaic market in Europe alongside Germany. In the period 2006–2018, Solar PV installed capacity grew at a CAGR of c. 66.3%, with an increase in 2011 versus 2010 y-o-y of 267.7% (Fig. 6).

Investments in renewable energy sources and the goal of contributing to combating climate change may also be considered as an opportunity for economic growth. The banking system has had a decisive role in supporting investments in renewables.

Intesa Sanpaolo, in particular, in 2018 alone, financed with €1922 million the green economy (over €18 billion between 2010 and 2018). The banking Group has been very active in the renewables sector in Italy, mainly in the solar space, financing historically over 20% of the market. The bank has been pivotal both as M&A advisor and financing bank, in the establishment and consolidation of some of the main solar players such as EF Solare, Tages and NextEnergy Capital.

In addition, the Group was supportive with most of the Italian renewables' champions such as Enel Green Power, Falck Renewables, ERG, and F2i throughout the years.

Case study: Interview to Andrea Mayr (former head of Investment Banking and Structured Finance at Banca IMI) on Enel Green Power (EGP) IPO.

(1) *How was Enel Green Power's IPO conceived?*
The Company was established two years before the IPO with the spin-off of the non-programmable renewables of Enel and the further addition of the renewables of Endesa. The rationale for the transaction was to unlock the hidden value of the perimeter of renewables, while keeping the control of the company.
(2) *What was Banca IMI's role in the IPO?*
Banca IMI acted as a Joint Global Co-Ordinator.
(3) *What was the distinctive positioning of EGP's equity story?*
When EGP approached the market, it was a global leader with a unique energy mix that included also geothermal and hydro capacity, while the competitors were more focused on wind. Moreover, the company was geographically well diversified, scarcely dependent on incentives and was benefitting from a strong and sustainable cash flow, underpinned by a higher load factor compared to its peers.
The vision of the top-management was anticipating the decreasing cost of technology and the need to evolve the renewables business models versus grid parity.
(4) *What was the reaction of the market?*
The IPO was planned in a period that suffered an intense volatility. This volatility impacted the international institutional demand, while on the contrary the Italian institutional and retail investors appreciated the transaction thanks to the very well-known reputation of the company. Intesa Sanpaolo had a pivotal role to attract the demand thanks to its network.
Originally it was expected a 70% Institutional Offer and 30% Retail Offer, but given the very high demand from private investors it was decided to change the structure of the Offer, focusing it more on retail investors. In the end, the latter filled 75% of the total demand and institutional investors 25%.

4 M&A as a Tool to Redefine the Structure of the Sector

With the ascent of sustainability in the global energy landscape there has been a marked change of course in the strategy of a multitude of companies. In some cases, this meant an adaptation of the business model, in others a radical and definitive departure from the previous corporate philosophy. In both circumstances, being the nature of the energy sector very competitive, one method more than any other has been chosen to drive the trend: M&A.

A perfect example of corporate audacity is represented by ERG SpA, which in less than a decade managed to shift from being one of the most important Italian oil companies, to become one of the most prominent European producers of energy generated from renewable sources.[1] According to top executive management of the Group, the idea was at first to become a multi-energy company, but soon the awareness that the growth of renewable sources would prove to be unstoppable and irreversible took over. A more drastic change was needed, one that would allow to focus on a less volatile sector (as Luca Bettonte, CEO of ERG, stated in a press conference) and anticipate the competition at the same time. Thus, it was decided to gradually abandon the oil sector in favor of an exclusive corporate focus on renewable sources, and the only way this was possible in such a swift way was via M&A.

It all started in 2006 with the acquisition of a controlling interest in EnerTAD SpA, a listed company engaged in the production of electricity from renewable sources. What followed next was a major transformation as, through dozens of transactions, ERG had in 2015 c. 73% of its Ebitda coming from renewables, up from circa 4% in 2008. This of course, was mainly feasible through M&A activities: in fact, in the same period, the Group was able to invest more than €3.5 billion in the production of renewable energy thanks to the gains from the disposal of its oil assets. Today, ERG has not only completely relocated from the oil sector but has also expanded beyond its historical borders of Italy, adapting to the more globalized nature of the world thanks to strategic acquisitions.

In this regard, one of the most striking Italian examples in which the impact of M&A on corporate culture is most visible is that of Enel SpA, in origin an all-Italian utility, now an international energy champion.[2] In the early 2000s the company started its global expansion project through several M&A operations: in 2000 it expanded into the US market acquiring Chi Energy, the first independent producer of alternative energy in the region, and the following year it bought Viesgo, a Spanish subsidiary of Endesa, active in the production and distribution of electricity, thus expanding its name in Iberia and opening the gateway also to the entire South American market. The internationalization process continued with multiple other operations, most notably the acquisition in 2007 of Iberic utility Endesa, which ultimately created an energy behemoth present in 22 countries and second in Europe for installed capacity.

Despite the fact that Enel remains an Italian reality and keeps a strong Italian presence, the results achieved because of its change in strategy, promoting a more global approach, have been impressive: as of December 31st 2018, the Group's Italian share of total revenues was c. 38%, and its operations are now distributed among 34 countries around the world with total managed generation capacity of more than 89 GW.

[1]The ERG case is described in detail in a chapter of this book.

[2]See the Enel's CEO contribution to this book.

Entering a new market is not a foregone conclusion and as can be inferred from the above, the fastest and most efficient way to do so is to venture into M&A activities. Diversifying risk has historically been one of the main rationales for expanding geo-graphically and, in particular, Italy has been a country of interest for many foreign energy operators because of its strong and developed consumer economy.

EDF, a French-based electricity producer, marketer and distributor, is one of the latter corporates that entered the Italian market. Being predominantly a nuclear electricity generator and distributor in France, the group wished to diversify its portfolio, both geographically and in terms of business model, and saw in Edison SpA, a major Italian gas company and Italy's second-biggest power provider at the time, the appropriate opportunity to achieve its objectives.[3]

In 2005 Transalpina di Energia Srl (TdE), the Italy-based joint venture between Delmi SpA and EDF, bought a controlling majority in Edison and in 2012 the Group increased its holding in Edison first by acquiring the remaining 50% stake in TdE and then by buying up to c. 99.5% of Edison's shares in a mandatory tender offer subsequent to the transaction. The fact that EDF already held a partial shareholding in Edison, in conjunction with an Italian utilities landscape characterized by a limited contenders' ability to challenge the financial power of the French Group, made the latter the favorite party to secure the operation.

The above represents only a handful of motivations and examples of the reasons for the implementation of inorganic growth strategies. However, in an increasingly competitive market, in which small businesses cannot withstand the price pressure of the biggest players and in which it becomes a necessity to constantly adapt the business model in order to retain and expand the customer base, M&A will play an even more crucial role in corporates' strategic planning than it did in the past. Intesa Sanpaolo, through Banca IMI, its investment banking arm, have played a pivotal role in shaping the Italian M&A environment and culture in the recent decades, providing strategic and financial advice to both domestic and international companies. Indeed, Banca IMI ranks 1st for number of transactions completed in Italy in the 2009–2018 period (300 deals) with a deal value of over €110 billion. In addition to the cases referred to above, Intesa Sanpaolo advised ERG in the acquisition of a 100% stake in Italy's hydroelectric business from E.On of 527 MW, Enel in the disposal of different Italian assets in order to finance its global expansion, and Edison in several M&A transactions completed in the last 15 years.

---

[3]See Edison contribution to this book.

5 Financial Investor's Strategies

However, utilities and energy companies were not the only parties active in the Italian energy market, and more specifically in renewables. A big role was played, and is still played, by financial investors. There are two types of renewable projects: brownfield and greenfield. So far, financial investors have concentrated more on the former, as it involves less risk and less technical know-how and ensures more immediate results.

Because of the above-mentioned incentives guaranteed by the Italian Government, renewables projects have historically provided the owner with a steady flow of earnings, so much so that some of the most involved financial investors are infrastructure funds, which normally seek long-term stability. The Italian renewables panorama has always been more attractive in terms of returns in comparison to other central and northern European countries, such as Norway, Sweden, France, the UK and Germany and this led, approximately one decade ago, financial investors to take a massive interest in the Italian energy market. Moreover, independent of government incentives, Italian market power prices have also been historically higher than the ones in the more Northern regions, a trend that is still actual.

A significant part of renewable installations now existing in Italy derive from the above-mentioned wave of investments and are in the hands of a handful of players. The market is however still very fragmented for the remaining portion of brownfield projects, which are sold at a higher price than what would be the cost of building a plant from the ground (greenfield), due to the incorporation of the much higher incentives these old installations still benefit from.

Nowadays, apart from the 4.8 GW of large-scale capacity that will be incentivized through the auction mechanism, a large part of the new installations will operate in grid-parity, selling electricity through Power Purchase Agreements (PPA). In fact, since the levelized cost of energy (LCOE) of solar PV and wind technologies has been in free-fall for years, and is forecasted to descend even more, there are more opportunities for greenfield project seekers, who will not require governmental incentives to breakeven, to enter into PPAs with offtakers, e.g. utilities or corporates.

Once the remaining brownfield projects are consolidated into bigger players, therefore, a new greenfield market will open up. Some financial investors will not acquire these projects in their first stages, but only enter in a second phase, once the construction is over. Nevertheless, we also expect that several players will seek agreements with developers, in order to secure access to the assets, avoiding the competitive sale processes, and to achieve better returns going up in the value chain.

The most active investment funds in Italy include F2i (Fondi Italiani per le Infrastrutture), NextEnergy and Tages, with which Intesa Sanpaolo holds a strong relationship proved by a series of recent collaborations in renewable portfolio investments, as visible in Fig. 7.

Companies	Date	Description	Banca IMI's role
Tages; Glennmont Partners	2018	Acquisition by Tages of Glennmont's PV portfolio (85.4MW)	M&A Advisor
Enel; EF Solare Italia; F2i Fondi italiani per le infrastrutture	2018	Enel's disposal of a 50% stake in EF Solare Italia to F2i	M&A Advisor
RTR; F2i Fondi italiani per le infrastrutture	2018	Acquisition by F2i of a 100% stake in the solar portfolio from RTR	M&A Advisor
RTR; PFI Awards 2018; F2i Fondi italiani per le infrastrutture	2018	Bridge to Equity. Acquisition Financing and refinancing of existing indebtedness €300m and €995m	MLA, Bookrunner, Agent Bank, Hedging Bank and Account Bank
Tages	2018	Refinancing of the indebtedness of a 139 MW PV portfolio (of which 101 MW already owned and 38 MW to be acquired) - €452.5m	MLA, Facility Agent
EF Solare Italia	2018	Existing indebtedness refinancing/new acquisition financing	MLA, Bookrunner, Hedging Bank and Original Lender
NextEnergy Capital	2017	Warehouse facility for future acquisitions in the Italian solar PV sector	Sole Lender
Gruppo Industriale Maccaferri; Tages	2017	Disposal by Maccaferri Group of a 51.2 MW PV portfolio to Tages	M&A Advisor
F2i Fondi italiani per le infrastrutture	2017	Existing indebtedness refinancing	MLA, Lender, Agent
Tages	2017	Refinancing of the existing indebtedness of a 101 MW photovoltaic portfolio	MLA
F2i Fondi italiani per le infrastrutture; Edison	2014	Acquisition by F2i of a 70% stake in a 600MW energy portfolio from Edison	M&A Advisor

Fig. 7 Selected Intesa Sanpaolo/Banca IMI's deals involving Italian financial investors. *Source* Intesa Sanpaolo

6 Climate Change and Sustainable Finance Development

On December 12th 2015, 196 governments entered into a multilateral agreement, the so-called Paris Agreement, aimed at tackling climate change, limiting global temperature increases to 1.5 °C above pre-industrial levels. The Paris Agreement's backbone is the so called "National Determined Contribution" ("NDC") framework, which sets the targets that each party voluntarily commits to achieve in order to reduce national emissions and adapt to the impacts of climate change. According to the Paris Agreement, the NDCs are revised every five years and potentially toughened, in order to reach the targets stipulated in the Paris accord.

Since then, great attention has been devoted to the climate change problem from governments, institutions, economic actors and public opinion. Nevertheless, all the efforts that have been put in place are still not enough. The latest data from the International Energy Agency (IEA) shows that energy-related CO_2 emissions grew 1.7% in 2018 from 2017 and that after two decades of continued growth, net renewables capacity additions in 2018 were equivalent to the ones in 2017.

According to the Global Climate Analysis report, published by the Carbon Disclosure Project (CDP) in June 2019 and based on the analysis of the world's largest companies (totaling US$17 trillion market capitalization), climate-related opportunities might amount to approximately US$2.1 trillion, driven by a potential increase in revenues due to the demand for low emissions products and services or better competitive positioning in the context of changes in consumer preferences. Financial companies forecast US$1.2 trillion in potential revenue from low emissions products and services. According to the OECD, in order to reach the Sustainable Development Goals set in 2015 and the targets of the Paris Agreement, nearly US$6.9 trillion a year of sustainable investments are needed by 2030.

Sustainable Finance will be one of the key enablers in the fight against climate change, as clearly identified during the last Climate Action Summit promoted in September 2019 in New York by UN Secretary General António Guterres, to enhance climate action. According to a joint UNDP/UNFCCC analysis, access to or availability of finance is the most critical factor, in particular for developing countries, in order to reach their NDCs.

In 2018, Green Finance global volumes were over US$170 billion: while green loans amounted to nearly US$60 billion (see Fig. 8), green bond issuances reached US$112.5 billion (see Fig. 9).

Green bonds, green loans and, especially, financing projects linked to renewable energies represent a clear focus for all European banks. In addition to the product offering, many of the latter are including Environmental Social and Governance ("ESG") aspects in their lending decisions and risk policies, and are creating dedicated ESG teams and committees, as well as introducing ESG reporting standards.

Fig. 8 Global green loans (US$ billion). *Source* Bloomberg

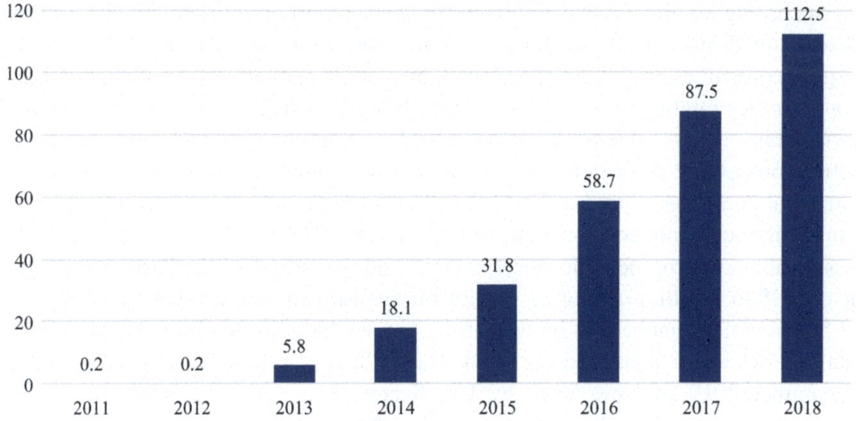

Fig. 9 Global green bonds (US$ billion). *Source* Refinitiv

Equity and debt investors are demonstrating a growing interest for sustainable finance, as they incorporate ESG targets in their investment strategies and develop specific products aimed at addressing climate change; in fact, 80% of investors already have ESG investment policies in place. Rating agencies as well are introducing methodologies to evaluate the ESG profile of an issuer, shifting attention from the sustainability of a single project—or a number of specific projects—to the sustainability of the whole corporate activities and strategy.

In 2018, the European Commission set up a Technical Expert Group on sustainable finance (TEG), involving representatives from civil society, academia, business and the financial sector, that on June 18th 2019 published reports and guidelines related to four key issues:

- EU classification system (the so-called EU Taxonomy) report, containing (i) technical screening criteria for 67 activities that can impact climate change mitigation (included in the sectors of agriculture, forestry, manufacturing, energy, transportation, water and waste, ICT and buildings), (ii) a methodology for evaluating the contribution to the climate change adaptation, (iii) guidance and case studies for investors;
- EU Green Bond Standard report, that proposes that the Commission create a voluntary, non-legislative EU Green Bond standard, to enhance the green bond market and foster market participation of issuers and investors;
- EU Climate benchmarks and benchmarks' ESG disclosures report, that set out the criteria for defining ESG indices;
- Guidelines on the disclosure of environmental and social information reports, that set out the guidelines to disclose relevant non-financial information in a consistent framework (Fig. 10).

Intesa Sanpaolo is dedicated to climate change issues with the awareness that innovation, the development of new products and services and corporate

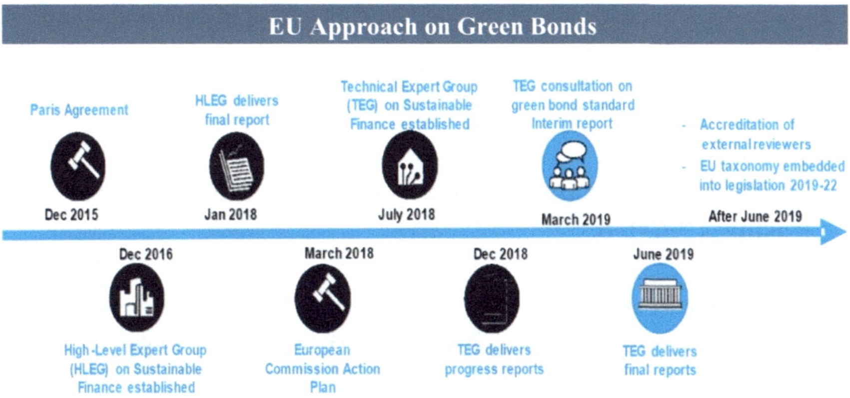

Fig. 10 EU's approach on green bonds. *Source* Banca IMI's DCM department

responsibility may contribute to tackle environmental changes and the related social impacts. This understanding has led over the years to adhere to numerous international standards, including the UNEP FI and the UN Global Compact, aimed at integrating environmental and social considerations into business operations. Furthermore, in October 2018 Intesa Sanpaolo decided to support the recommendations of the Task Force on Climate-related Financial Disclosures (TCFD), with the voluntary commitment to disseminate transparent reporting on the risks and opportunities linked to climate change.

For nine years in a row Intesa Sanpaolo has been included—as the only Italian bank—in the Dow Jones Sustainability Index Europe and Dow Jones Sustainability Index World, one of the most prominent global and European financial market indexes for sustainability. This background, together with the strong drive towards innovation within Intesa Sanpaolo, led to the creation of an Innovation Center, that among its many objectives, is committed to the development of the circular economy, a new paradigm aimed at decoupling economic development from the exploitation of finite natural resources and redesigning the industrial system.

In 2018, the Bank's commitment to the circular economy, developed in partnership with the Ellen MacArthur Foundation, was reinforced with the establishment of a credit plafond of €5 billion and the launch, together with Fondazione Cariplo, of the first Circular Economy Lab in Italy, dedicated to the circular economy and the businesses that intend to adopt this approach.

Intesa Sanpaolo's credit process considers, among others, environmental and social risks. The Group is continuously enhancing its internal processes, to apply more stringent assessment criteria and more effective operating procedures. Since 2007, Intesa Sanpaolo has been focusing on project finance initiatives, in which environmental and social risks are assessed through the Equator Principles guidelines. Over the last 12 years, a total of 346 loans have been subjected to screening for the Equator Principles, and in 2018, more than 20 projects, totaling over €990 million, were deemed compliant with such principles and financed by the Bank.

In June 2017, Intesa Sanpaolo was the first Italian bank to issue a €500 million Senior Unsecured Green Bond, connected with environmental sustainability projects.

7 Challenges in Financing the Energy Transition and New Business Models

The 2030 targets set by the Paris Agreement are significantly reshaping the industry structure and the business models. Those targets, embedded in the Clean energy for all Europeans package, envisage the decrease of CO_2 emissions by up to 30% versus 2005 (33% in Italy), the increase of renewable capacity by up to 32% of the final energy consumption (30% in Italy) and the reduction of the primary energy consumption compared to the scenario Primes 2007 by −32.5% (43% in Italy).

Let us look at the power sector, as an example of the disruptive trends that are reshaping the energy markets. The traditional paradigm was based on a centralized electricity production and distributed consumption. Power generation was characteristically made by large power plants (mainly nuclear, thermal or hydroelectric power plants) and electricity was delivered to the final consumer (industrial, residential and public services' entities) through the transmission and distributions networks. In the current paradigm the consumer model is evolving into a prosumer model, with the spreading of small and distributed power plants (mainly solar installations) that will transform the final customer into both a consumer and a producer.

The increase of non-programmable renewables (i.e. wind and solar depending on the atmospheric conditions), will create instability in the networks. In Italy electricity generation units grew from 700 to over 1 million, and further growth is expected to several million more, driven by the decarbonization targets. Solar PV was around 4 GW in 2010 and reached 20.1 GW in 2018. It should almost triple in terms of installed capacity reaching 50.9 GW in 2030 (8.0% CAGR 2018–2030). Similarly, wind energy was c. 5 GW in 2010 and reached 10.3 GW in 2018. It should almost double in terms of installed capacity to c. 18.4 GW in 2030 (4.9% CAGR 2018–2030). In parallel, the Italian thermoelectric capacity decreased of 20 GW from 2012 to 2018, and the reserve margin was reduced from 25 to 6 GW, an amount equal to the Italian import capacity.

The electricity system is being impacted by the introduction of battery storage technology. Energy storage deployment is expected to grow at a rate of 40%, mainly because the technology is dramatically improving, extending the life cycle and reducing the production costs. The storage will represent an important complement to renewables' development, allowing to store the electricity for delivery when needed, and representing an important element for the flexibility of the networks.

Another important trend that is reshaping the power sector is the electrification of consumptions. In particular, the electrification of the transportation sector could have a relevant impact, although it is doubtful that it will become mainstream in the near term. According to Bloomberg New Energy Finance, over 2 million electric vehicles were sold in 2018, up from just a few thousand in 2010. And it is expected that more than 50% of all passenger vehicle sales will be electric by 2040 (and around 30% of the global passenger vehicle fleet). Nevertheless, internal combustion engine-cars still represent 99% of the total global car population. The story telling of the need to reduce emissions is not enough to convince the mass market. The limited extension of the EV charging networks, the lengthy charge time compared to refueling a tank and the high cost of EVs still represent limits for their diffusion. However, the growth of EVs might create further impact on the electricity system, due to the batteries that withdraw electricity from the network but might also feed-in electricity when they are charged, with the additional complication of the vehicle's moving location.

The evolution of market structures and of players' business models creates many challenges for the financial market and the banking system. A significant effort needs to be made, in order to analyze and assess the technology's stability, the new revenue models and the potential stranded assets.

A further important trend, already mentioned, is that the Italian renewables market will mainly develop with grid-parity projects. In the first months of 2018, more than 2.5 GW of PV subsidy-free capacity has been announced across Portugal, Spain, Italy and France. Subsidy-free projects generate income without receiving any government support. Without public subsidies, developers need to reduce the volatility of captured power prices in order to be able to finance project construction through long-term PPAs. The financing of a renewable project with a long term PPA envisages a different approach compared to an incentivized plant and need among the many elements an assessment of the offtaker credit capacity, of the pricing structure of the PPA and of the tenor of the contract. Another very important factor is to regulate the event of termination of the PPA, protecting the risk profile of the project.

In conclusion, many disruptive trends will reshape the energy sector, and the challenges for the banks will be to stay at the forefront of the energy transition, together with the energy players, in order to support them to succeed in this remarkable evolution of the sector.

The manufacturer's authorised representative in the EU is Springer Nature Customer Service Centre GmbH, Europaplatz 3, 69115 Heidelberg, Germany. If you have any concerns regarding our products, please contact ProductSafety@springernature.com

Printed and bound by CPI Group (UK) Ltd, Croydon, CR0 4YY
30/09/2026
02225602-0009